Dye Chemistry - Exploring Colour From Nature to Lab

Edited by Brajesh Kumar

Published in London, United Kingdom

Dye Chemistry - Exploring Colour From Nature to Lab
http://dx.doi.org/10.5772/intechopen.1001671
Edited by Brajesh Kumar

Contributors
Alexey V. Nizovtsev, Birkishore Mahto, Deepti Pargai, Dirk Labudde, Eugene A. Tikhonov, Ghulam Sabir Hussain, Girish Chandra, Gopal Kumar Mahato, Hussaini Majiya, Igor P. Ilchishin, Janet Joshiba Ganesan, Jia-Ren Chang Chien, Joshua Inanabor Isibor, Kazi Md. Rashedul Islam, Kirill Kolmakov, Laura Pistorius, M. Abdul Jalil, M. Mahbubul Bashar, Massimiliano Lucidi, Paul Seidel, Ronny Bodach, Shahin Hossain, Sidra Mubeen, Svenja Preuß, Tahira Batool, Tarikul Islam, Tommy Bergmann, Ujala Rani

First published in London, United Kingdom, 2024 by IntechOpen
IntechOpen is the global imprint of INTECHOPEN LIMITED, registered in England and Wales, registration number: 11086078, 167-169 Great Portland Street, London, W1W 5PF, United Kingdom

British Library Cataloguing-in-Publication Data
A catalogue record for this book is available from the British Library

Additional hard and PDF copies can be obtained from orders@intechopen.com

Dye Chemistry - Exploring Colour From Nature to Lab
Edited by Brajesh Kumar
p. cm.
Print ISBN 978-1-83769-685-7
Online ISBN 978-1-83769-684-0
eBook (PDF) ISBN 978-1-83769-686-4

For EU product safety concerns:
IN TECH d.o.o., Prolaz Marije Krucifikse Kozulić 3, 51000 Rijeka, Croatia,
info@intechopen.com or visit our website at intechopen.com.

Meet the editor

Dr. Brajesh Kumar is an assistant professor and the head of the Department of Chemistry, TATA College, Kolhan University, Chaibasa, India. He received a Ph.D. in Chemistry from the University of Delhi, India, in 2009. He is a pioneering researcher in the fields of green chemistry, nanosynthesis, environmental remediation, photocatalysis, nanomedicine, and polymer and natural product extraction. Dr. Kumar has several national and international fellowships to his credit and has worked as a faculty member at various universities throughout India, Ecuador, and South Korea. He is a holder of two registered patents as well as a member of the American Chemical Society, Indian Science Congress Association (ISCA), and the Indian Society of Chemists and Biologists (ISCB). He has published more than eighty research articles and is an active reviewer of more than seventy journals. He was also listed among the World's Top 2% of Scientists by Stanford University, USA, in 2021, 2022, and 2023.

Contents

Preface

Dyes and colors have been integral to human culture for centuries, used by ancient civilizations such as India, China, Egypt, Rome, and Greece to decorate various materials in daily life. The color of molecules is generally attributed to electronic transitions between π-type molecular orbitals in the ultraviolet/visible (UV/Vis) region, and these transitions are influenced by factors such as the number and position of substituents, double bonds, conjugation length, chromophore structure, degree of oxidation, size, and charge. Synthetic colorants can be produced through full chemical synthesis or by modifying natural precursor compounds like indigoids, curcumin, carotenoids, xanthophylls, betanine, anthocyanins, and chlorophylls. Natural colorants are derived from sources such as plants, animals, microorganisms, and minerals. While dyes have long been used in industries like textiles, wood, leather, metals, ceramics, drugs, cosmetics, energy, and food, there are ongoing debates about their potential health effects.

This book is a collection of research and review papers on the chemistry and physics of dyes and their intermediates, covering such topics as chemical constituents, spectroscopic aspects, surface, solution, crystal formation, photochemistry, and their ecological and biological properties.

In Chapter 1, Pargai summarizes the general chemistry and functional properties of natural dyes. Chapter 2, by Islam et al., sheds light on the diverse sources of natural dyes, their fastness properties, and current efforts to enhance their durability, paving the way for a more vibrant and sustainable future in the textile industry. In Chapter 3, Majiya discusses new perspectives, challenges, and prospects for using food-grade dye extracts for disinfecting foods and contact surfaces. In Chapter 4, Mubeen et al. highlight the potential of waste onion (Allium cepa) peels, rich in phytochemicals like tannins, quercetin, and pelargonidin, as a natural dyestuff for textiles, supporting green chemistry principles.

Chapter 5, by Chandra et al., reviews the recent literature on azobenzene, a well-known dye that undergoes fast trans-cis photoisomerization and emerging applications in multifunctional sensing. In Chapter 6, Tikhonov and Ilchishin study and develop laser systems using dye-doped matrices in nematic and cholesteric liquid crystals (LCs), discussing the lasing properties of these lasers with tunable wavelength oscillations at recording flashy reflective Bragg gain gratings of variable spatial periods in crossed pumping beams. In Chapter 7, Kolmakov et al. focus on the development of rhodamine dye derivatives as fluorophores for stimulated emission depletion (STED) microscopy, emphasizing the factors affecting imaging quality, photostability, and reactive label stability.

Chapter 8, by Bergmann et al., demonstrates the role of color analysis in forensics, demonstrating how colors can support forensic methods to draw meaningful conclusions from recorded data. They emphasize the importance of continued advancements in color analysis technologies and methods, as well as the establishment of standards and best practices to further support crime-solving efforts. In Chapter 9, Isibor explains the synthesis of some azo dyes using aniline compounds and different coupling agents, with the synthesized dyes exhibiting good light and wash fastness properties on cellulosic fabrics.

The final chapter, by Chang Chien and Joshiba, examines adsorbents for dye degradation, the mechanisms and effectiveness of the adsorption process, and recommendations for adopting clean and cost-effective water purification practices, especially given the environmental and health risks posed by dye-contaminated wastewater.

I would like to express my deep gratitude to my beloved wife, Kumari Smita, for her insightful comments on several chapters and her invaluable support in conceptualizing the book's content.

Dr. Brajesh Kumar
Department of Chemistry,
TATA College,
Kolhan University,
Jharkhand, India

Section 1

Natural Dye and Applications

Chapter 1

Chemistry of Natural Dye for Functional Applications

Deepti Pargai

Abstract

Natural dyes offer a fascinating blend of esthetics and functionality. Their chemical structures and interactions with fabrics can provide functional properties like UV protection, bacterial resistance, fungal resistance, mosquito repellence, and flame retardancy. Besides these specific functional properties, natural dyes also offer different functions such as moisture management, antioxidant effects, thermoregulation, insect-repellent properties, biocompatibility, and aromatherapy which not only provide comfort to the wearer but also provide health benefits. This chapter tries to explore the chemistry behind these. This chapter explores the chemistry of natural dyes in textile applications. This chapter also provides insights into the extraction and application process for the application of natural dyes on textiles. Standard Testing procedures for testing the functional properties as well as testing procedures for the presence of chemical bonds such as FTIR are also mentioned in the chapter for better understanding and references. The Chemistry of Natural dye is a very dynamic and vast topic and exploration of it in context to functional properties will surely help to visualize the future of natural dyes.

Keywords: FTIR, functional natural dyes, application process of natural dyes, chemistry of natural dye, targeted dye design, future of natural dyes

1. Introduction

The captivating world of natural dyes offers more than just vibrant hues. Hidden within their complex chemical structures lies a treasure trove of functional properties with immense potential for the textile industry. The chemistry of natural dyes is fascinating and complex. It plays a vital and very significant role in their functional properties. Functional properties such as UV protection, bacterial resistance, fungal resistance, mosquito repellence, flame retardancy, and odor control can be attributed to the presence of specific bioactive compounds in the dye source [1–5]. Some natural dyes, such as alizarin and curcumin, exhibit antibacterial activity [6]. Certain natural dyes like flavonoids or polyphenols can absorb UV radiation, offering protection from the sun's harmful rays [7]. By understanding the chemistry of natural dyes, their potential for various functional applications can be harnessed. The future of natural dyes can be promising as these dyes have the ability not only to provide color on textiles with vibrant and colorfast shades but also can enhance comfort and health-related benefits to the wearer. This chapter tries to explore the interaction between the dye's chemical structure, the dye-fabric bond, and the resulting functional effect. This collaborative approach between chemists and textile scientists paves the way for the development of novel, sustainable textiles with enhanced properties.

In a traditional way, the color of a natural dye is primarily determined by the presence of specific functional groups in its molecule. These functional groups, known as chromophores, absorb certain wavelengths of light and reflect others [8]. The reflected wavelengths correspond to the color we perceive. For example, the anthraquinone group is responsible for the red color of alizarin [9]. Similarly, the indigoid group is responsible for the blue color of indigo. Understanding the chemistry of natural dyes in textile applications is essential for optimizing their functional performance and ensuring product quality. Methods for testing chemical bonds on fabric, such as FTIR, as well as evaluating functional properties using standard test methods provide valuable insights into the efficacy of natural dye treatments. Incorporating FTIR analysis allows for detailed characterization of dye-fabric interactions, aiding in the development of sustainable and functional textiles with enhanced performance and consumer appeal [10].

Although synthetic dyes can also provide these functional properties [11, 12] to the fabric in addition to their functional properties, natural dyes offer several sustainable advantages in textile applications. These are biodegradable and environmentally friendly, minimizing the environmental impact of textile manufacturing and disposal. Unlike some synthetic dyes that may contain harmful chemicals, natural dyes are generally non-toxic and safe for use, making them suitable for sensitive skin and reducing health risks for workers in the textile industry. Many natural dyes have been used for centuries in traditional textile crafts, preserving cultural and artistic heritage while supporting sustainable practices. Textiles dyed with natural dyes contribute to environmental conservation by reducing water pollution and minimizing the release of harmful chemicals into ecosystems. Natural dyeing processes often require less water and energy compared to conventional synthetic dyeing methods, leading to lower carbon emissions and a smaller ecological footprint.

Besides the social and environmental advantages, natural dyes can also contribute some additional comfort as well as health-related beneficial properties to textiles. Natural dyes can contribute to improved moisture management properties in textiles. Some natural fibers, such as cotton and wool, have inherent moisture-wicking abilities, and when dyed with natural dyes, these properties can be further enhanced. This is particularly beneficial in activewear and outdoor textiles, where moisture management is crucial for comfort and performance [13–15]. Certain natural dyes contain antioxidant compounds that can impart antioxidant properties to textiles. For example, polyphenolic compounds found in plant-based dyes like turmeric or tea extracts have antioxidant effects. Fabrics dyed with these natural dyes may offer additional protection against oxidative damage, potentially extending the lifespan of the textile and providing benefits for the skin [16]. However, it needs further in-depth exploration in the case of textiles as it is very difficult to maintain that kind of antioxidant effect through it. Some natural dyes possess insect-repellent properties. Fabrics dyed with these natural dyes can help deter insects such as mosquitoes, ticks, or mites, providing added protection for outdoor or camping textiles [17]. Natural dyes are often biocompatible, meaning they are compatible with the human body and less likely to cause allergic reactions or skin irritations [18]. This makes textiles dyed with natural dyes suitable for sensitive skin or applications where direct contact with the skin is required, such as baby clothing or medical textiles.

Some natural dyes derived from aromatic plants contain volatile compounds with aromatic or therapeutic properties. Fabrics dyed with these natural dyes may retain subtle fragrances or therapeutic benefits, contributing to a sensory experience and potential mood enhancement for the wearer [19, 20]. Despite so many advantages, the durability of natural dyes on the textile surface is a major challenge. However, experimenting and having knowledge of the chemistry between textile and dye will surely help to enhance the durability aspect of dyes.

2. Chemical structure responsible for color as well as functional properties

The coloration properties of natural dyes are heavily influenced by their chemical structures. Various classes of natural dyes exhibit distinct color characteristics based on their molecular compositions. Flavonoids, such as quercetin and kaempferol, feature conjugated systems of double bonds and hydroxyl groups, resulting in yellow to orange hues. Indigoid dyes, derived from plants like *Indigofera tinctoria*, produce a range of blue shades. Anthraquinone dyes, typified by compounds with anthraquinone structures, offer red colors and boast good light fastness. Alpha-hydroxy-naphthoquinones, exemplified by lawsone or henna from *Lawsonia inermis*, contribute to natural yellow hues. Additionally, anthocyanins found in fruits and flowers exhibit pH-dependent color variations, transitioning from red in acidic conditions to blue or purple in neutral or alkaline environments **Table 1**. Carotenoids, featuring long conjugated double bonds, contribute to the yellow to red hues observed in various plant sources. The chemical structures of these natural dyes dictate their absorption spectra and color properties, making them valuable for diverse applications in textiles, food, and cosmetics industries (**Figure 1**) [34].

Class	Color	Chemical Structure	Name of the Dye	Source	Ref.
Indigoid	Deep Blue	Indigo molecule (fused indole and indigo ring)	Indigo	*Indigofera tinctoria, Isatis tinctoria*	[21]
Anthraquinone	Reds, Purples	Anthraquinone core (two benzene rings linked by a carbonyl group)	Madder	*Rubia tinctorum*	[22, 23]
Anthraquinone	Reds	Anthraquinone core	Lac	*Laccifer lacca (resin)*	[24, 25]
Flavonoid (Quercetin)	Yellow	Phenolic rings with various substitutions	Onion	Onion peels	[26, 27]
Flavonoid (Myricetin)	Yellow-Gold	Phenolic rings with various substitutions	*Madhuca longifolia*	*Madhuca longifolia* leaves	[28]
Flavonoid	Yellow	Phenolic rings with various substitutions	Luteolin	Dyer's weld (*Reseda luteola*)	[29, 30]
Carotenoid	Yellow, Orange	Long chain with alternating single and double bonds, Crocin	Saffron	Saffron (*Crocus sativus*)	[31, 32]
Other	Yellow	*Curcuminoids*	Turmeric	Curcuma longa	[33]

Table 1.
Chemical structure of common natural dyes [21–34].

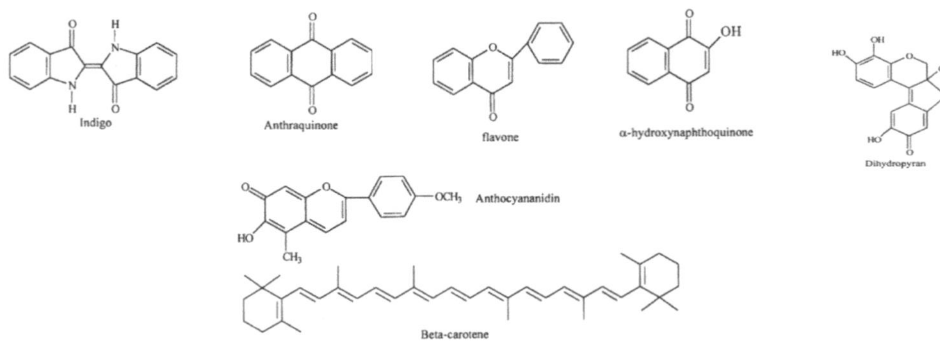

Figure 1.
Chemical structure of the natural dye [21–34].

3. Chromophores and Auxochromes responsible for color and functional properties

Chromophores are responsible for the absorption of light by natural dyes, leading to coloration. These chromophores often consist of conjugated pi-electron systems, such as double bonds or aromatic rings, which absorb specific wavelengths of light. Auxochromes, on the other hand, are functional groups that enhance the color intensity and stability of natural dyes by extending the conjugation of the chromophores or facilitating bonding with the substrate.

3.1 Chromophores

Chromophores are chemical groups or structural features within molecules that are responsible for the absorption of light, thereby giving rise to color. These groups typically contain conjugated pi-electron systems, such as alternating single and double bonds or aromatic rings. Conjugation refers to the arrangement of alternating single and double bonds or aromatic rings along a molecule's backbone, which allows for the delocalization of electrons. The delocalization of pi-electrons results in the absorption of specific wavelengths of light from the visible spectrum, leading to the manifestation of color. Common examples of chromophores include double bond and aromatic rings.

Double bonds include conjugated systems of double bonds, such as those found in the carbonyl group (-C=O) or in polyenes (chains of alternating single and double bonds), which can absorb light and contribute to coloration. For instance, the beta-carotene molecule contains a long chain of conjugated double bonds, giving rise to its orange color [35, 36]. This can also be responsible for UV protection.

Aromatic rings include aromatic compounds, such as benzene rings, which can also act as chromophores. The delocalized pi-electrons within the ring system absorb light, leading to coloration. For example, anthocyanins in berries and flowers contain aromatic rings that contribute to their red, purple, or blue hues. This can be responsible for functional properties such as UV protection and antibacterial properties [37].

3.2 Auxochromes

Auxochromes are functional groups attached to a chromophore that enhance the color intensity and stability of the dye molecule. These functional groups typically contain lone pairs of electrons that can interact with the chromophore, thereby extending the conjugation of pi-electrons or facilitating bonding with the substrate. Auxochromes can alter the electronic structure of the chromophore, which affects its absorption spectrum and color appearance. Examples of common auxochromes are hydro, amino, and carboxy groups. Hydroxyl groups are common auxochromes that can donate lone pairs of electrons to the chromophore, extending its conjugation and enhancing its color intensity. For example, hydroxyl groups attached to flavonoid molecules, such as quercetin and kaempferol, contribute to their yellow to orange hues. Amino groups can act as auxochromes by donating lone pairs of electrons to the chromophore, leading to enhanced color intensity. For instance, the amino groups in the indigo molecule contribute to its deep blue coloration. Carbonyl groups, such as those found in ketones and aldehydes, can act as auxochromes by participating in resonance structures with the adjacent chromophore, thus influencing its color properties [38, 39].

4. pH sensitivity of natural dyes

Many natural dyes exhibit pH-dependent color changes, a phenomenon known as pH sensitivity. This property arises from the presence of weakly acidic or basic functional groups within the dye molecules. For example, anthocyanins can appear red, purple, or blue depending on the acidity of the environment due to reversible protonation/deprotonation of hydroxyl or carbonyl groups. Many natural dyes contain functional groups such as hydroxyl (-OH), carboxyl (-COOH), or amino (-NH2) groups that can act as weak acids or bases. These functional groups can undergo reversible protonation (gain of a hydrogen ion, H^+) or deprotonation (loss of a hydrogen ion) reactions depending on the pH of the surrounding medium [40].

In acidic conditions, the equilibrium shifts toward the protonated form of the dye molecule, which typically exhibits a different absorption spectrum and color compared to the deprotonated form. Conversely, in alkaline conditions, deprotonation occurs, leading to a different molecular structure with altered electronic properties and thus a different color. The pH sensitivity of natural dyes arises from the influence of pH on the distribution of protonated and deprotonated forms of the dye molecule, which affects its absorption properties and color appearance [41].

pH sensitivity is essential in various applications such as food coloring, cosmetics, and textile dyeing, where precise control over color appearance is required. By adjusting the pH of the dyeing solution or the substrate, different colors can be achieved using the same dye molecule. Understanding pH sensitivity allows for the development of pH-responsive materials and color-changing indicators for analytical and diagnostic purposes.

For example, anthocyanins are a class of water-soluble pigments found in various fruits, vegetables, and flowers. They are known for their pH-dependent color changes. Anthocyanins typically exist in different forms depending on the pH of the solution. In acidic conditions (lower pH), they are predominantly in the flavylium cation form, which appears red. As the pH increases toward neutral or alkaline conditions, anthocyanins undergo structural transformations into other forms such as quinoidal bases and chalcones, resulting in colors ranging from purple to blue. The color changes observed in anthocyanins are due to reversible protonation and deprotonation of specific functional groups within the molecule, such as hydroxyl (-OH) or carbonyl (-CO) groups [42, 43].

5. Process of natural dye for functional application

5.1 Extraction of natural dyes

Natural dyes can be extracted using different mediums such as water alcohol or steam using different extraction methods i.e., maceration, exhaustion, Soxhlet extraction, Microwave extraction, UV radiation, etc.

Medium for extraction can be different as water, alcohol (ethanol methanol), ether, etc. Many natural dyes can be extracted using water as a solvent. This involves boiling or soaking the dye source such as plant materials {leaves, roots, fruit, and vegetable peels}, insects, or minerals in water to release the colorants. Some dye sources are better soluble in alcohol or alcohol-water mixtures. Alcohol extraction is often used for dyes with low water solubility or for extracting specific compounds. Steam distillation is employed for extracting volatile natural dyes, such as essential oils from plant materials.

5.2 Mordanting

After extraction of dyes, mordanting process is done in the case of pre- and simultaneous mordanting methods. Mordants are substances used to fix natural dyes to fibers and improve colorfastness. Common mordants include alum, iron, copper, tin, and chrome. Mordanting can be done in different stages: pre-mordanting, simultaneous mordanting, and post-mordanting. In pre-mordanting, fibers are treated with mordants before dyeing. This ensures better absorption and fixation of the dye. In Simultaneous mordanting, the dyeing and mordanting process is done at the same time in a single dye bath. Post-mordanting involves treating dyed fibers with mordants after dyeing. It allows for the modification of color shades and enhances colorfastness.

5.3 Dyeing parameters

Concentration: The concentration of dye solution and mordant solution affects the intensity and depth of color. Higher concentrations generally result in darker shades.

Temperature: Dyeing temperature influences the rate of dye uptake and fixation. Different dyes and fibers have optimal dyeing temperatures.

Time Range

Extraction: Extraction time varies depending on the dye source and method. It can range from a few minutes (for some plant materials) to hours (for certain roots, barks, or wood).

Dyeing: Dyeing time depends on factors such as fiber type, dye concentration, and desired color intensity. It typically ranges from 30 minutes to a few hours.

Exhaustion: Exhaustion refers to the point at which the dye bath becomes nearly colorless, indicating that most of the dye has been absorbed by the fibers. Exhaustion time varies depending on dye concentration, fiber type, and mordanting process. Exhaustion occurs when the dye bath is depleted of dye, indicating that the dye has been absorbed by the fibers [44–46].

Effects of parameters on functional properties: These parameters not only affect the color of the dye but also affect the functional properties of the fabric. It has been reported in several studies that the UPF of the fabric increases with the increment of the concentration of dye.

6. Functional properties of natural dyes: chemistry in action

6.1 UV protection

Our understanding of the chemistry behind natural dyes' UV protection properties is continuously evolving. We know that conjugated double bonds present in dyes like curcumin (turmeric) and carotenoids effectively absorb harmful UV rays. Additionally, anthraquinones (madder root, buckthorn berries) with their aromatic ring structures contribute to UV shielding.

If we talk about dye-fabric bonds, a strong covalent or ionic bond formed with mordants (like alum) is crucial. Research demonstrates how different mordants like alum, ferrous sulfate, and cupric sulfate can influence the UV protection rating of fabrics dyed with alizarin.

By delving deeper into the relationship between specific chemical structures and UV absorption, chemists can design natural dyes with enhanced UV protection. This could involve modifying existing dyes or exploring new natural sources rich in UV-absorbent molecules.

Textile scientists are investigating how mordants, beyond strengthening the dye-fabric bond, can influence UV protection. Exploring synergistic effects between different mordant types and specific dye molecules could lead to optimized UV-protective textiles [1, 47, 48].

6.2 Bacterial protection

Natural dyes possess a fascinating arsenal for combating bacteria and fungi. Phenolic hydroxyl groups (juglone and lawsone) disrupt bacterial membranes, while tannins (oak bark, chestnut) can inhibit fungal growth by binding to essential proteins and enzymes.

Phenolic hydroxyl groups are not the only warriors against bacteria. Natural dyes rich in quinones, like juglone (from walnut husks) and lawsone (from henna), exhibit strong antibacterial activity. Quinones can disrupt bacterial cell membranes and even generate reactive oxygen species that damage bacterial cells.

Similar to UV protection, a strong bond with the fabric is essential. A study explored the antibacterial properties of cotton fabrics dyed with various natural dyes and found that a strong dye-fabric bond, achieved through mordanting with tannic acid, enhanced the antibacterial effect against *E. coli* [49, 50].

6.3 Mosquito repellence

Volatile terpenoids are not the only mosquito-repellent weapons in nature. Essential oils like citronella oil, derived from citronella grass, contain a range of active ingredients, including citronellal and limonene. These compounds can disrupt mosquito odor receptors and feeding behavior. While a strong fabric bond is still beneficial for long-lasting repellency, some repellency can come from loosely bound dye molecules containing volatile terpenoids. These can create a mosquito-unfriendly zone around the fabric through vapor release [51].

6.4 Flame retardancy

Natural dyes like those extracted from pomegranate peels or grape seeds have shown promise as flame retardants for textiles. These dyes contain compounds that can decompose upon exposure to heat, releasing non-flammable gases and forming a char layer that slows down flame propagation. Some natural dyes can act as flame retardants, such as banana peduncle extract, safflower yellow derivative, and spinach leaves juice. For example, banana peduncle extract contains high levels of nitrogen, calcium, sodium, potassium, phosphorus, iron, and magnesium, which may make it a flame retardant. Safflower yellow derivative can be used to create a dyed, flame-retardant, and UV-resistant silk fabric. Spinach leaves juice can also be used to give cotton fabric flame-retardant properties [52, 53].

6.5 Odor reduction

Some natural dyes possess odor-absorbing properties. Fabrics dyed with chlorophyll (from green plants) or chitosan (from crustacean shells) have been shown to exhibit some level of odor reduction, potentially due to their ability to bind odor-causing molecules [54].

7. Standard test methods to test functional properties

7.1 Antimicrobial activity

Natural dyes with inherent antimicrobial properties can be evaluated for their effectiveness in inhibiting microbial growth on textiles. Standardized methods such as AATCC Test Method 100 or ASTM E2149 assess the antimicrobial activity of treated fabrics against bacteria or fungi [54, 55]. UPF measures the effectiveness of textiles in blocking harmful UV radiation. Testing involves spectrophotometric analysis to determine the UV transmittance of dyed fabrics across different wavelengths, allowing the calculation of UPF values according to established standards (e.g., AATCC Test Method 183 or ASTM D6603) [56, 57]. Natural dyes with mosquito-repellent properties can be assessed using cone bioassays or cage tests. Fabric treated with mosquito-repellent dyes can be evaluated for its ability to deter mosquito landing and feeding, quantifying repellence efficacy through established protocols [58]. IS 2004 test method is the determination of flammability and flame resistance of textile fabric [59].

7.2 FTIR identification of functional group in dyes

FTIR analysis allows researchers to identify the functional groups present in natural dyes, such as hydroxyl (-OH), carbonyl (C=O), and aromatic (C=C) groups. By comparing the FTIR spectra of natural dyes to reference spectra or databases, researchers can determine the chemical composition and structural characteristics of the dye molecules. FTIR spectroscopy works based on the principle of absorption of infrared radiation by chemical bonds within a sample. When infrared light passes through a sample, certain wavelengths are absorbed by specific chemical bonds, causing them to vibrate. The resulting absorption spectrum provides information about the types of chemical bonds present in the sample and their molecular environments [60]. Peak assignments and intensity changes in FTIR spectra can indicate the formation of chemical bonds, crosslinking, or interactions between dye molecules and fabric substrates.

7.3 Application of FTIR in the field of natural dyes

FTIR spectroscopy can be used to identify specific functional groups present in plant-based dyes. For example, the presence of characteristic absorption bands corresponding to flavonoids or anthocyanins can help distinguish between different plant species used for dye extraction. Insect-based dyes, such as cochineal extract obtained from the cochineal insect, contain unique compounds like carminic acid. FTIR analysis can reveal the presence of carboxylic acid groups (COOH), characteristic of carminic acid, aiding in the identification and quality control of cochineal-based dyes. Certain natural dyes, such as ochre or malachite, are derived from mineral sources. FTIR spectroscopy can detect characteristic absorption bands associated with metal-oxygen (M-O) vibrations in these mineral pigments, allowing for their identification and differentiation. FTIR analysis can also be used to monitor the dyeing process in real time. By analyzing changes in the FTIR spectra of the dye solution and the dyed material at different stages of the process, researchers can assess dye uptake, binding interactions, and dye stability [61, 62].

Interpretation of FTIR spectra involves identifying characteristic absorption bands corresponding to specific functional groups. Peak positions, intensities, and shapes provide information about the chemical environment and interactions of the functional groups within the dye molecules. Software tools and databases are often used to assist in the interpretation and comparison of FTIR spectra for accurate identification and characterization of natural dyes (**Table 2**).

8. Analysis of functional properties through FTIR

By identifying functional groups like aromatic rings with conjugated double bonds in the dye molecules, FTIR can indicate potential UV absorption properties. These groups are often associated with UV protection capabilities. The presence of certain functional groups, such as amides, phenols, or carboxylic acids, in the dye molecule might suggest potential antibacterial activity. However, further tests are needed to confirm this activity. Similar to antibacterial properties, the presence of specific functional groups, like amides, terpenes, or certain aromatic structures, can be indicative of potential insect-repellent properties. However, this needs to be confirmed with bioassays specifically designed for insect repellency evaluation.

The effectiveness of these functionalities also depends on the concentration and type of natural dye used, as well as how well it bonds to the fabric. In conclusion, FTIR is a valuable tool for identifying the functional groups present in natural dyes on fabrics, which can provide indications of potential UV protection, antibacterial activity, and insect repellence. However, for confirmation of these properties, further tests specific to each function are essential.

Dye (Common Name)	Class	UV Vis absorption Spectrum	FTIR (approx. range)	Bonds	References
Curcuma Longa (Haldi, Indian Saffron): Root	Curcumin	420–580 nm	maximum at 1625 cm^{-1}	The presence of an aromatic structure, C=C group C=O	[63, 64]
Indigo: leaves	Indigoid	375–475 nm	1500–1600 cm^{-1}, 1340–1470 cm^{-1}, 1180–1360 cm^{-1}	C=C, C-H, C-N	[65, 66]
Beetroot: root	Betalains	Strong absorption band at 530	1056 cm(−1) with a baseline established between 1187 and 887 cm(−1).	O-H, C=O and C-O-C	[67, 68]
Lawsonia Inermis (Henna, Mehndi): leaves	2-hydroxy-1, 4-naphthoquinone	673 nm	broad peak between 3650 and 3076 cm^{-1}. The peak at 2927 and 2852 cm^{-1} … … further peaks at 1734 and 1637 cm^{-1}	corresponding to the stretching vibrations of the -O-H group corresponding to aliphatic -C-H group corresponding to -C=O.	[69, 70]
Madder: root (Manjistha)	Anthraquinone	398–426 nm	3600–3200 cm^{-1}, 1640–1620 cm^{-1}, and 1590–1570 cm^{-1}.	O=H, C=O, C=C	[71]
Marigold: flowers	Carotenoid (Lutein and Luten)	242–250 nm	1640–1620 cm^{-1}, 1590–1570 cm^{-1}, 1500–1900 cm^{-1}	C=O and C=C and carbonyl group	[71]

Table 2.
UV Vis absorption, FTIR, and prevalent Bonds of some common natural dyes.

9. Challenges

Scalability is a major challenge for natural dyes. While natural dyes offer environmental benefits, scaling up production for large-scale textile manufacturing could be a challenge. Research efforts focused on cost-effective extraction techniques and high-yielding natural dye sources are essential. While some brands are incorporating natural dyes, large-scale adoption remains a challenge. Scaling up production of natural dyes to meet the demands of major clothing companies requires further research and development.

Natural dyes can exhibit variations in color and functionality due to factors like plant growth conditions and extraction methods. Developing standardized extraction and application procedures is crucial for consistent results in textile production. Standardized process is not available for each dye. Natural dyes can exhibit variations in color and functionality due to factors like plant growth conditions and extraction methods. Developing standardized processes for consistent results is crucial for commercial viability. Lack of awareness about the benefits of natural dyes is also a major concern. Lack of awareness means less demand leads to less supply and less research in this field. Raising awareness about the benefits of functional natural dyes can encourage consumer demand and support the growth of this sustainable alternative.

9.1 Opportunities and the future of functional natural dyes

The beauty of natural dyes lies in their ability to combine esthetics with functionality. By harnessing the power of chemistry, we can unlock the full potential of these natural colorants for a more sustainable and healthy future. Research in natural dye chemistry is ongoing. By understanding the specific functionalities offered by different dye molecules and their interaction with fabrics, scientists can develop targeted applications. This could lead to the creation of textiles with built-in UV protection for outdoor clothing, antibacterial properties for medical textiles, or even mosquito-repellent clothing for travel in high-risk areas. As research progresses and production methods improve, we can expect to see wider adoption of functional natural dyes in commercial applications. Collaborations between scientists, textile manufacturers, and fashion brands will be key to unlocking the full potential of this exciting field.

Bioactive Dye Discovery is being done. Chemists are actively searching for novel natural dyes with potent antimicrobial properties. This could involve investigating lesser-known plant extracts or utilizing bioprospecting techniques to identify new sources with specific functionalities. Patagonia and some other international brands have started to promote functional properties like sun-protective clothing. But still could not offer a complete range of natural dyes clothes having functional properties. Textile scientists are exploring methods to improve the durability of the dye-fabric bond, particularly for different functionalities. This could involve advancements in mordant chemistry or the development of new fixation techniques.

10. Conclusion

The world of natural dyes is a treasure trove waiting to be explored. Understanding the chemistry behind these functionalities is crucial for optimizing their application in textiles. This research brings together the expertise of chemists and textile scientists to explore this exciting realm. With a collaborative spirit and a focus on innovation, we can unlock the full potential of natural dyes for a sustainable and functional future. The future of functional natural dyes lies in a collaborative approach between chemists and textile scientists. Chemists can explore the vast library of natural products, identifying novel dye sources rich in specific functionalities. Textile scientists, in turn, can develop innovative dyeing and mordanting techniques to optimize dye-fabric interactions and ensure durability. By combining expertise, we can unlock the full potential of natural dyes, creating a new era of sustainable and functional textiles that not only look beautiful but also offer built-in protection and enhanced performance. Fourier Transform Infrared (FTIR) spectroscopy is a powerful analytical technique used for the characterization of natural dyes. It provides detailed information about the molecular structure and functional groups present in the dye molecules. By unraveling the intricate relationship between their chemistry and functional properties, we can revolutionize the textile industry.

Author details

Deepti Pargai
Textile Design, Mody University of Science and Technology, Sikar, Rajasthan, India

*Address all correspondence to: pargai.deepti16@gmail.com

References

[1] Shahidi S, Khoshechin E, DalalSharifi S, Mongkholrattanasit R. Investigation of the effect of various natural dyes on UV protection properties and antibacterial activity of cotton fabrics. Journal of Natural Fibers. 2022;**19**(13):7213-7228. DOI: 10.1080/15440478.2021.1944433

[2] Tegegne W, Haile A, Zeleke Y, Temesgen Y, Bantie H, Biyable S. Natural dyeing and anti-bacterial finishing of cotton fabric with extracts from Justicia schimperiana leaf extract: A step towards sustainable dyeing and finishing. International Journal of Sustainable Engineering. 2024;**17**(1):1-10. DOI: 10.1080/19397038.2023.2301702

[3] Jahan N, Arju SN. A sustainable approach to study on antimicrobial and mosquito repellency properties of silk fabric dyed with neem (*Azadirachta indica*) leaves extractions. Sustainability. 2022;**14**(22):15071. DOI: 10.3390/su142215071

[4] Bar G, Bar M. Dyeing and flame retardant finishing of silk fabric: An ecofriendly approach. SN Applied Sciences. 2020;**2**:363. DOI: 10.1007/s42452-020-2114-8

[5] Begum A, Al Mamun A, Khan MA, Adnan AM. Developing odor free textile with natural colorants using new nontoxic mordant. International Journal of Scientific & Engineering Research. 2017;**8**(12):1949-1957. ISSN 2229-5518

[6] Balakrishnan N, Ostheller M, Aldeghi N, Schmitz C, Groten R, Seide G. Pilot-scale electrospinning of PLA using biobased dyes as multifunctional additives. Polymers. 2022;**14**:2989. DOI: 10.3390/polym14152989

[7] Li L et al. Natural products and extracts from plants as natural UV filters for sunscreens: A review. Animal Models and Experimental Medicine. 2023;**6**(3):183-195. DOI: 10.1002/ame2.12295

[8] Colour sensation. 2024. Available from: https://webstor.srmist.edu.in/web_assets/srm_mainsite/files/2017/Dyes-Notes.pdf

[9] Dulo B, Phan K, Githaiga J, et al. Natural Quinone dyes: A review on structure, extraction techniques, analysis and application potential. Waste and Biomass Valorization. 2021;**12**:6339-6374. DOI: 10.1007/s12649-021-01443-9

[10] Priyadharsini P, Theodore David S, Joel. Natural dyes and their FT-IR spectroscopy studies. International Journal on Applied Bioengineering. 2014;**8**(2)

[11] Noser AA, Ibrahim SA, Abd El Salam HA, et al. Pyrazole-vaniline Schiff base disperse azo dyes for UV protective clothing: Synthesis, characterization, comparative study of UPF, dyeing properties and potent antimicrobial activity. Journal of the Iranian Chemical Society. 2023;**20**:2963-2976. DOI: 10.1007/s13738-023-02891-x

[12] Ibrahim SA, Selim AI, Sakr AM, et al. Design, synthesis, characterization, and antimicrobial properties of new azo disperse dyes incorporating Quinazolinone-Pyrazolone moieties and their applications for dyeing of polyester fabrics. Fibers and Polymers. 2024;**14**(3):3-13. DOI: 10.1007/s12221-024-00507-6

[13] Tutak M, Onal L, Benli H. Investigation of wicking properties of naturally-dyed linen fabrics. Tekstil ve Konfeksiyon. 2013;**23**:374-380

[14] Ullaha MH, Akthera H, Rahmand MM, Foisalb ABM, Hasanc MM, Zumahid SMA-A, et al. Surface modification and improvements of wicking properties and dyeability of grey jute-cotton blended fabrics using low-pressure glow discharge air plasma. Heliyon. 2021;**7**:e07893

[15] Dave H, Ledwani L, Chandwani N, Kikani P, Desai B, Chowdhuri MB, et al. Use of dielectric barrier discharge in air for surface modification of polyester substrate to confer durable wettability and enhance dye uptake with natural dye eco-alizarin. Composite Interfaces. 2012;**19**(3-4):219-229. DOI: 10.1080/15685543.2012.702594

[16] Ángeles Bonet-Aracil M, Díaz-García P, Bou-Belda E, Sebastiá N, Montoro A, Rodrigo R. UV protection from cotton fabrics dyed with different tea extracts. Dyes and Pigments. 2016;**134**:448-452. ISSN 0143-7208. DOI: 10.1016/j.dyepig.2016.07.045. Available from: https://www.sciencedirect.com/science/article/pii/S014372081630362X

[17] Singh A, Sheikh J. Synthesis of a novel cationic dye to impart mosquito-repellent and UV protection to an acrylic fabric. ACS Omega. 2023;**8**(11):10214-10224. DOI: 10.1021/acsomega.2c07693

[18] Costa EM et al. Insights into the biocompatibility and biological potential of a chitosan nanoencapsulated textile dye. International Journal of Molecular Sciences. 2022;**23**(22):14234. DOI: 10.3390/ijms232214234

[19] Mehta S, Macgillivray M. Analysis of stress relief effect by use of aroma-therapeutic textiles. AATCC Journal of Research. 2016;**3**(5):38-44. DOI: 10.14504/ajr.3.5.4

[20] Hilarylhahn. Dye-and-aromatherapy-workshop. 2024. Available from: https://www.hilarylhahn.com/blog/botanical-dye-and-aromatherapy-workshopon2024

[21] Mocquard J, Le Lamer A-C, Fabre P-L, Mathieu C, Chastrette C, Vitrai A, et al. Indigo dyeing from *Isatis tinctoria L.*: From medieval to modern use. Dyes and Pigments. 2022;**207**:110675. ISSN 0143-7208. DOI: 10.1016/j.dyepig.2022.110675. Available from: https://www.sciencedirect.com/science/article/pii/S0143720822005976

[22] Fouillaud M, Caro Y, Venkatachalam M, Grondin I, Dufossé L, Nollet alML, et al. Phenolic Compounds in Food Characterization and Analysis. U.K: CRC Press; 2018. pp. 130-170. DOI: 10.1201/97813151201579. 978-1-4987-2296-4. hal-01657104

[23] Derksen GCH. Analysis and isolation of anthraquinones from madder roots (*Rubia tinctorum*). Wageningen Universiteit. 2001. Article ID 546839. DOI: 10.1155/2013/546839

[24] De Santis D, Moresi M. Production of alizarin extracts from *Rubia tinctorum* and assessment of their dyeing properties. Industrial Crops and Products. 2007;**26**:151-162

[25] Wei B, Chen Q-Y, Chen G, Tang R-C, Zhang J. Adsorption properties of lac dyes on wool, silk, and nylon. 2013;**2013**:1-6

[26] Hollman PCH, van Trijp JMP, Buysman MNCP, Gaag MSVD, Mengelers MJB, de Vries JHM, et al. Relative bioavailability of the antioxidant flavonoid quercetin from various foods in man. FEBS Letters. 1997;**418**(1-2):152-156. ISSN 0014-5793. DOI: 10.1016/S0014-5793(97)01367. Available from: https://www.sciencedirect.com/science/article/pii/S0014579397013677

[27] Pucciarini L, Ianni F, Petesse V, Pellati F, Brighenti V, Volpi C, et al. Onion (*Allium cepa L.*) skin: A rich resource of biomolecules for the sustainable production of Colored biofunctional textiles. Molecules. 2019;**24**(3):634. DOI: 10.3390/molecules24030634

[28] Swamy VN, Gowda KNN, Sudhaka R. Dyeing of silk using *Madhuca longifolia* as natural dye source. Indian Journal of Fibre & Textile Research. 2015;**40**:419-424. Available from: https://www.researchgate.net/publication/286934192_Dyeing_of_silk_using_Madhuca_longifolia_as_natural_dye_source [Accessed: Mar 31, 2024]

[29] Vankar PS, Shukla D. Spectrum of colors from reseda luteola and other natural yellow dyes. Indian Journal of Textile Engineering & Fashion Technology. 2018;**4**(2):105-118

[30] Ellis C. Natural dye: Experiments and results a blog by Catharine Ellis weld (Reseda luteola) 2015. Available from: https://blog.ellistextiles.com/2015/10/02/weld-reseda-luteola/

[31] Gutheil WG et al. Crocetin: An agent derived from saffron for prevention and therapy for cancer. Current Pharmaceutical Biotechnology. 2012;**13**(1):173-179. DOI: 10.2174/138920112798868566

[32] Laura S, Fabio G, Giuseppe R. Saffron (*Crocus Sativus L.*) apocarotenoids: A review of their biomolecular features and biological activity perspectives. Carotenoids: Properties, Effects and Diseases. 2011:145-178

[33] Kocaadam B, Şanlier N. Curcumin, an active component of turmeric (*Curcuma longa*), and its effects on health. Critical Reviews in Food Science and Nutrition. 2017;**57**(13):2889-2895. DOI: 10.1080/10408398.2015.1077195

[34] Vankar PS. Chemistry of natural dyes. Resonance. 2000;**1**:73-80

[35] Kennepohl D, Farmer S, Soderberg T, Reusch W. Conjugation, color, and the chemistry of vision. Libretex Chemistry. Jan 2024. Available from: https://chem.libretexts.org/Bookshelves/Organic_Chemistry/Organic_Chemistry_(Morsch_et_al.)/14%3A_Conjugated_Compounds_and_Ultraviolet_Spectroscopy/14.09%3A_Conjugation_Color_and_the_Chemistry_of_Vision#:~:text=Beta%2Dcarotene%20has%2011%20conjugated,the%20complementary%20color%20of%20blue

[36] da Silva Oliveira JP, de Oliveira RT, Guedes AL, da Costa Oliveira M, Macedo AF. Metabolomic studies of anthocyanins in fruits by means of a liquid chromatography coupled to mass spectrometry workflow. Current Plant Biology. 2022;**32**:100260. ISSN 2214-6628. DOI: 10.1016/j.cpb.2022.100260. Available from: https://www.sciencedirect.com/science/article/pii/S2214662822000275

[37] Yuzuak S, Xie D-Y. Anthocyanins from muscadine (*Vitis rotundifolia*) grape fruit. Current Plant Biology. 2022;**30**:100243. DOI: 10.1016/j.cpb.2022.100243

[38] Vishwanathan M. Principles of Organic Chemistry. Kollam, Kerala: Jai Sai Publications. 1922. pp. 2-30

[39] Chequer FMD, de Oliveira GAR, Ferraz ERA, Cardoso JC, Zanoni MVB, de Oliveira DP. Textile Dyes: Dyeing Process and Environmental Impact. U.K: Intechopen; 2019

[40] Kumar, Gupta V. Fundamentals of Natural Dyes and its Application on Textile Substrates. London, UK:

IntechOpen; 2020;**2019**(1):1-32. DOI: 10.5772/intechopen.89964

[41] Van der Schueren L, De Clerck K. Coloration and application of pH-sensitive dyes on textile materials. Coloration Technology. April 2012;**128**(2):82-90. DOI: 10.1111/j.1478-4408.2011.00361.x

[42] Enaru B et al. Anthocyanins: Factors affecting their stability and degradation. Antioxidants (Basel, Switzerland). 2021;**10**(12):1967. DOI: 10.3390/antiox10121967

[43] Khoo HE et al. Anthocyanidins and anthocyanins: Colored pigments as food, pharmaceutical ingredients, and the potential health benefits. Food & Nutrition Research. 2017;**61**(1):1361779. DOI: 10.1080/16546628.2017.1361779

[44] Ragab MM, Hassabo AG, Othman HA. An overview of natural dyes extraction techniques for valuable utilization on textile fabrics. Journal of Textiles Coloration and Polymer Science. 2022;**19**(2):137-153

[45] Sayem ANM, Ahmed F, Saha P, Talukder B. A review on natural dyes: Raw materials, extraction process, and their properties. Advance Research in Textile Engineering. 2021;**6**(1):1062

[46] Karpagam Chinnammal S. Textile colouration and finishes. In: Prakash, editor. Chapter: 9 Natural Mordants. Xinhe: Hebei Guangxing Chemical Industry Co., Ltd; 2024. Available from: https://ebooks.inflibnet.ac.in/hsp09/chapter/natural-mordants/#:~:text=Simultaneous%20mordanting:%20In%20simultaneous%20mordanting%2C%20the%20material,washed%20properly%20and%20soaping%20is%20carried%20out

[47] Feng XX, Zhang LL, Chen JY, Zhang JC. New insights into solar UV-protective properties of natural dye. Journal of Cleaner Production. 2007;**15**(4):366-372. ISSN 0959-6526. DOI: 10.1016/j.jclepro.2005.11.003

[48] Grifoni D, Bacci L, Zipoli G, Albanese L, Sabatini F. The role of natural dyes in the UV protection of fabrics made of vegetable fibres. Dyes and Pigments. 2011;**91**(3):279-285. ISSN 0143-7208. DOI: 10.1016/j.dyepig.2011.04.006

[49] Singh R, Jain A, Panwar S, Gupta D, Khare SK. Antimicrobial activity of some natural dyes. Dyes and Pigments. 2005;**66**(2):99-102. ISSN 0143-7208. DOI: 10.1016/j.dyepig.2004.09.005

[50] Pargai D, Jahan S, Gahlot M. Functional Properties of Natural Dyed Textiles. London, UK: IntechOpen; 2020. DOI: 10.5772/intechopen.88933

[51] Kantheti P, Rajitha I, Padma A. Natural finishes on textiles to combat the mosquitoes: A pilot study. Journal of Entomology and Zoology Studies. 2020;**8**(2):30-33

[52] Faysal G, Kowser M, Jalil M. Recent development of sustainable ecological flame-retardant textile composite material: A review. Journal of Materials Science and Chemical Engineering. 2022;**10**(6):30-44. DOI: 10.4236/msce.2022.106004

[53] Basak S, Samanta K, Chattopadhyay S. Fire retardant property of cotton fabric treated with herbal extract. The Journal of The Textile Institute. 2014;**106**:1-10. DOI: 10.1080/00405000.2014.995456

[54] Chowdhury A. Anti-odour finishing of textiles. Research Gate; 2008;**55**:88-94+96

[55] Microbes Investigations. 2024. Available from: https://microbe-investigations.com/aatcc-100/

[56] ASTM. Available from: https://www.astm.org/e2149-20.html. 2024

[57] Hatch KL. American standards for UV-protective textiles. In: Dummer R, Nestle FO, Burg G, editors. Cancers of the Skin. Recent Results in Cancer Research. Vol. 160. Berlin, Heidelberg: Springer; 2002. DOI: 10.1007/978-3-642-59410-6_6

[58] Al Parvez A, Hossain J, Hossain Z, Sohan MSH, Hoque F, Ahsan H, et al. Mosquito repellent fabric: Development and characterization of peppermint and garlic mixture finish on knitted fabric to examine mosquito repellency. Heliyon. 2023;**9**(5):e15944. ISSN 2405-8440. DOI: 10.1016/j.heliyon.2023.e15944

[59] IS 2004. Methods for determination of flammability and flame resistance of textile fabric. 2024

[60] Lee J, Kang MH, Lee K-B, Lee Y. Characterization of natural dyes and traditional Korean silk fabric by surface analytical techniques. Materials. 2013;**6**(5):2007-2025. DOI: 10.3390/ma6052007

[61] Samanta A. Bio-Dyes, Bio-Mordants and Bio-Finishes: Scientific Analysis for their Application on Textiles. 2020. DOI: 10.5772/intechopen.92601

[62] Setiawan T, Subekti W, Nur'Adya S, Khusnul I, Siti U. Performance of natural dye and counter electrode from Robusta coffee beans peel waste for fabrication of dye-sensitized solar cell (DSSC). IOP Conference Series: Materials Science and Engineering. 2018;**299**:012073. DOI: 10.1088/1757-899X/299/1/012073

[63] Sofyan N et al. Visible light absorption and photosensitizing characteristics of natural yellow 3 extracted from *Curcuma Longa L.* for dye-sensitized solar cell. IOP Conference Series: Earth and Environmental Science. 2018;**105**:012073

[64] Van Nong H, Hung LX, Thang PN, Chinh VD, Vu LV, Dung PT, et al. Fabrication and vibration characterization of curcumin extracted from turmeric (*Curcuma Longa*) rhizomes of the northern Vietnam. Springerplus. 2016;**5**(1):1147. DOI: 10.1186/s40064-016-2812-2

[65] Suyitno B, Kristiawan B. Absorbance and electrochemical properties of natural indigo dye. AIP Conference Proceedings. 2018;**1931**:030067. DOI: 10.1063/1.5024126

[66] Wahyuningsih S et al. Indigo dye derived from *Indigofera Tinctoria* as natural food colorant to cite this article. IOP Conference Series: Materials Science and Engineering. 2017;**193**:012048

[67] Devadiga D, Ahipa TN. Betanin: A Red-Violet Pigment - Chemistry and Applications. U.K: Intechopen; 2020. DOI: 10.5772/intechopen.88939

[68] Garrigues JM, Akssira M, Rambla FJ, Garrigues S, de la Guardia M. Direct ATR-FTIR determination of sucrose in beet root. Talanta. 2000;**51**(2):247-255. DOI: 10.1016/S0039-9140(99)00258-1. Available from: https://www.sciencedirect.com/science/article/pii/S0039914099002581

[69] Abeid OK, Zhao T, Seeahmed A. Dyeing and functional finishing of cotton fabric using henna extract and TiO2 nano-sol Dyeing and functional finishing of cotton fabric using henna extract and TiO2 nano-sol. Fibers and Polymers. 2015;**16**(6):1303-1311. Available from: https://www.researchgate.net/figure/ATR-FTIR-spectra-of-a-cationized-cotton-treated-with-TiO-2-nano-sol-and-dyed-with-Henna_fig2_282307416 [Accessed: May 15, 2024]

[70] Jamkai N, HKM, Suhaimiii S, Omarii SR. Optimization of dye extraction from turmeric (*Curcuma Longa L.*) and henna leaves (*Lawsonia inermis L.*) in dye sensitized solar cell. USIM Research Repository System. 2020. Available from: https://oarep.usim.edu.my/entities/publication/a18daf5e-5940-4302-8c87-8c8c72ae9e85/full

[71] Nong HV, Hung IX, Thang PN, Chnh VD, Vu LV, Pung PT, et al. Bio-dyes, bio-mordants and bio-finishes: Scientific analysis for their application on textile. In: Chemistry and Technology of Natural and Synthetic Dyes and Pigments. NY: Springer Plus; 2020. DOI: 10.5772/intechopen.92601

Chapter 2

Understanding the Fastness Issues of Natural Dyes

Tarikul Islam, Kazi Md. Rashedul Islam, Shahin Hossain, M. Abdul Jalil and M. Mahbubul Bashar

Abstract

The utilization of natural dyes in textile production has gained significant attention due to their eco-friendly characteristics and minimal environmental impact. Serving as a sustainable alternative for textile coloring, particularly when derived from native plant species, natural dyes contribute to the promotion of local biodiversity. Obtained from various botanical sources such as flora, flowers, leaves, roots, berries, barks, and wood, they offer a diverse range of hues spanning blues, reds, yellows, browns, and violets. However, achieving consistent and predictable colors with natural dyes presents challenges due to inherent variations in plant sources, growing conditions, and extraction techniques. The integration of mordants, including aluminum potassium sulfate, potassium dichromate, copper sulfate, and others, is crucial to enhance dye absorption and improve colorfastness. Natural dyes are employed across fabrics like cotton, linen, silk, wool, hemp, and blends, each contributing unique qualities to the vibrant and environmentally friendly palette. Yet, the longevity and strength of colors may vary based on factors like dye type, fabric substrate, and mordant effects, impacting chemical bonding between fibers, dyes, and mordants. Hence, meticulous selection of dyes and mordants, considering their compatibility with specific fibers, is essential for achieving optimal colorfastness and durability in natural dyeing processes.

Keywords: natural dyes, color fastness, color strength, mordants, sustainability

1. Introduction

Natural dyes, derived from plants, animals, minerals, and microorganisms, have been utilized for centuries to imbue textiles with vibrant hues. Classified based on their origin, including plant, animal, mineral, or microorganism dyes, natural pigments offer a renewable, biodegradable, and sustainable alternative to synthetic dyes. While plants predominantly serve as the primary source of natural dyes, their utilization extends across various realms, each offering unique properties and colors. The art of natural dyeing not only unveils the rich spectrum of colors hidden within local flora but also highlights the intricate relationship between humans and their environment. Through experimentation and collaboration, botanists and domestic gardeners have explored optimal cultivation practices, including soil conditions, climate suitability, and harvesting techniques, to maximize dye yields. This synergy between science and tradition has paved the way for the development of modern cultivation systems, ensuring the sustainable production of high-quality natural dyes. Contrary to common misconceptions, natural dyes are not limited to earthy tones but encompass a diverse array of vibrant and fast colors. With meticulous extraction processes and strategic dehydration methods, the inherent richness and luster of natural dyes rival that of synthetic counterparts, if not surpassing them. The commitment to harnessing the full potential of natural dye sources has fueled ongoing research and exploration, leading to the discovery of newer sources and expanding the color palette available for textile dyeing.

In recent years, there has been a resurgence of interest in natural dyes, particularly for dyeing cotton and silk fabrics [1]. This renewed focus has spurred the exploration of innovative techniques and novel sources, promising an array of newer shades and fabric options. As farmers and fabric companies delve deeper into the realm of natural dyes, they stand to unlock new opportunities for the production of ethnic fabrics and natural products, contributing to a more sustainable and environmentally conscious future [2]. Through this chapter, we aim to explore the intricate world of natural dyes, shedding light on their origins, properties, and cultivation practices. By delving into the art and science of natural dyeing, we seek to inspire further exploration and innovation in this timeless craft, ensuring its enduring legacy in the textile industry and beyond.

2. Source of natural dyes

The utilization of natural dyes in textile dyeing and printing dates back to the mid-nineteenth century, marking an era where synthetic dyes coexisted alongside their natural counterparts. Despite the economic efficiency and superior fastness properties of synthetic dyes, the introduction of natural dyes prompted a reduction in their usage [3]. The resurgence of natural dyes can be attributed to several factors, including increasing consumer awareness of the harmful effects associated with synthetic dyes, mounting global environmental concerns, and the implementation of stringent environmental regulations. This shift toward eco-conscious practices has propelled the revival of natural dyes, emphasizing their sustainability and biodegradability.

Natural dyes originate from various sources, including plants, biological organs, minerals, insects, and microbial and fungal sources, each offering a diverse spectrum of coloring matter. Despite this diversity, these sources can be broadly classified into four main categories based on their origin.

2.1 Plant source

In the annals of history, natural dyes predominantly emanated from plants, offering a plethora of hues sourced from various botanical materials such as roots, leaves, twigs, stems, bark, flowers, fruits, and more. This discussion delves into the realm of plant-derived natural dyes, showcasing their diverse range and traditional significance. Some plant sources of natural dyes are listed in **Table 1**.

Henna leaves, revered for their use in traditional hand painting, are a popular source of natural dye. Women often utilize henna leaves to create intricate designs on their hands. The process involves drying and crushing the leaves, followed by boiling them with water to extract the dye. The resulting hues span from rich browns to

DOI: http://dx.doi.org/10.5772/intechopen.1005363

Name	Biological name	Part used	Main component	Chemical structure	Color	Reference
Hena	*Lawsonia inermis*	Leaf	Lawsone	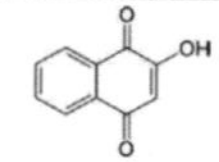	Brown to yellow	[4]
Tulsi	*Ocimum sancum*	Leaf	Ursolic acid	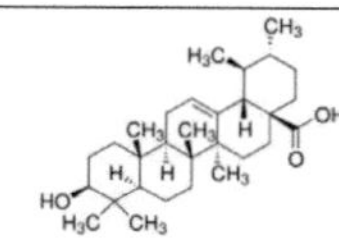	Green	[5]
Red sandalwood	*Pterocarpus santainus*	Wood	Santarubins	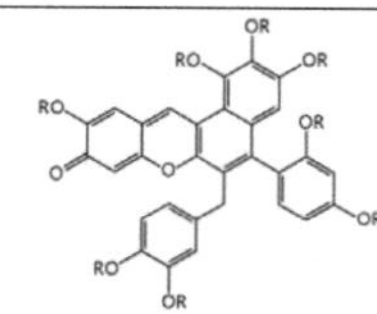	Red	[6]
Turmeric	*Curcuma longa*	Root	Curcumin	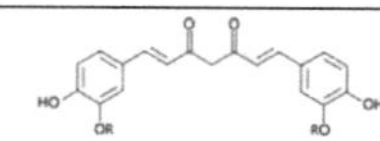	Yellow	[7]
Pomegranate fruit peel	*Punica granatum*	Peel	Tannin, flavonoids	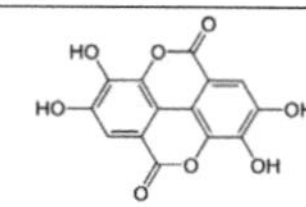	Green to yellow	[8]
Chamomile	*Matricaria recutia*	Flower	Flavonoids	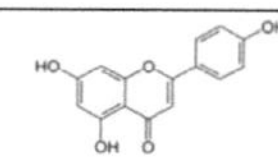	Green or yellow	[9]
Madder	*Rubia cordifolia*	Root and stem	Purpurin	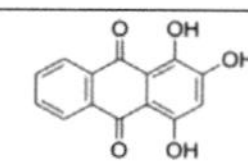	Brown	[10]

Name	Biological name	Part used	Main component	Chemical structure	Color	Reference
Onion outer shell	*Allium cepa*	Outer shell	Pelargonidin	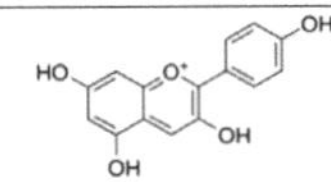	Orange to brown	[7]
Logwood	*Haematoxylum campechianum*	Leaf	Hematoxylin	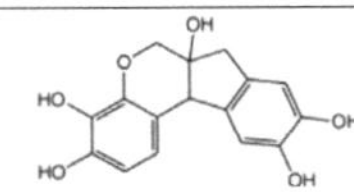	Black	
Saffron	*Crocus sativus*	flower	Kaempferol, Pelargonidin	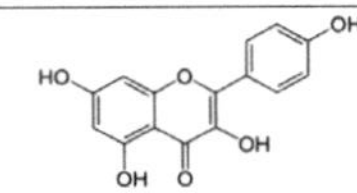	Yellow	

Table 1.
Plant source of natural dyes.

vibrant mustard yellows, adding a touch of cultural vibrancy to textiles. Tulsi leaves, enriched with ursolic acid, offer another botanical source of natural dye, yielding shades of verdant green. This versatile leaf serves not only as a medicinal herb but also as a sustainable option for textile dyeing, infusing fabrics with a fresh, botanical hue. Red sandalwood, renowned for its therapeutic properties, finds application as both an herbal remedy and a natural dye source. Extracted from the heartwood of the plant, red sandalwood yields a spectrum of colors ranging from warm oranges to deep browns, imbuing textiles with earthy, natural tones.

Turmeric, celebrated for its culinary and medicinal uses, also serves as a valuable natural dye. The roots of the turmeric plant are dried, ground into a fine powder, and boiled to extract the dye. Known as curcumins, the pigments in turmeric impart textiles with a radiant golden hue, evoking warmth and vitality. Pomegranate fruit peel extract offers a unique natural dye option, producing hues of green–yellow that transition to olive and dark gray when combined with iron. Additionally, pomegranate peel serves as a tannin-rich mordant, enhancing color fixation and durability on fabrics. Chamomile flowers, with their delicate blooms, yield natural dyes in shades of green or yellow. These flowers contain a variety of flavonoids, including apigenin, luteolin, and quercetin, which contribute to their color properties, adding a touch of natural elegance to textiles. Indian madder, a member of the Rubiaceae family, is valued for its roots and rhizomes, which produce anthraquinone pigments ideal for fabric dyeing. Whether used with or without mordants, Indian madder imparts textiles with hues ranging from warm oranges to rich browns, adding depth and character to fabrics. Logwood, indigenous to regions like Mexico and Central America, is prized for its ability to produce deep, rich black hues when used to dye wool, lending sophistication and depth to textiles. Saffron, extracted from the stigma of the flower, imparts a luminous yellow hue to fabrics, adding a touch of luxury and opulence to textiles, evoking warmth and vitality. Through the judicious utilization of these plant-based natural dyes, artisans and designers can create textiles imbued with rich, sustainable colors, while honoring age-old traditions and ecological principles [11].

2.2 Biological organ source

In ancient times, a variety of rich red and purple hues were derived from natural dyes sourced from animals. Among these, cochineal stands out, extracted from insects, and renowned for its vibrant crimson, scarlet, and pink tones. Cochineal dye can be enhanced through mordanting with substances like alum, chromium, iron, and copper, resulting in a spectrum of colors from brown to red. The carmine pigment, derived from female cochineal insects, contains high concentrations of carminic acid, ranging from 19% to 22%.

Additionally, lac dye, extracted from stick lac through a process involving water, sodium carbonate, and lime, yields a resinous substance that imparts reddish-purple hues, spanning from scarlet to crimson. Furthermore, murex snails, a species of snail, produce a distinct purple dye highly prized in ancient dyeing practices (see **Table 2**) [12].

2.3 Mineral source

Mineral dyes are naturally occurring earth pigments renowned for their tinctorial properties, primarily derived from oxides or hydrated oxides. Among the diverse array of mineral dyes are chrome yellow, iron buff, Prussian blue, nankin yellow, and manganese brown [17]. Found abundantly in nature, minerals such as red ochre,

Name	Biological name	Part used	Main component	Chemical structure	Color	Reference
Cochineal insecrs	*Dactylopius coccus*	Whole body	Carminic acid	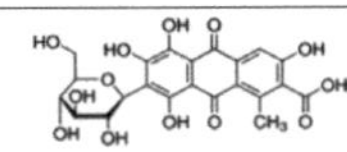	Brownish red	[12]
Lac insect	*Kerria lacca*	Whole body	Laccaic acid	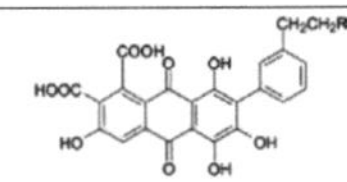	Red	[13]
Murex Snails	*Muricidae murex*	Shell	6,6 dibromoindigo	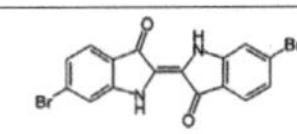	Purple	[14]
Polish Cochineal	*Porphyrophora polonica*		Carminic acid	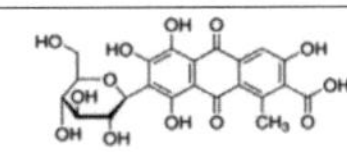	Red	[15]
Octopus	*Octopus vulgaris*	Ink	Melanin	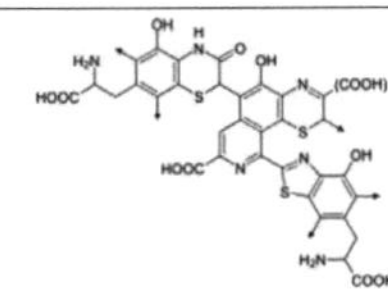	Black	[16]

Table 2.
Biological source of natural dyes.

Dyes name	Chemical structure	Extract color	Reference
Ochre	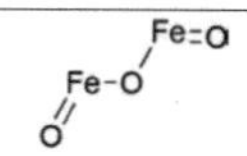	Yellow, brown, red nuances	[18]
Manganese	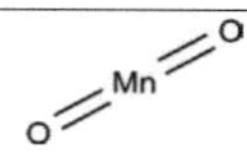	Black nuances	[19]
Cinnabar	Hg = S	Red nuances	[20]
Azurite	$Cu_3(CO_3)_2(OH)_2$	Blue nuances	[21]
Aragonite	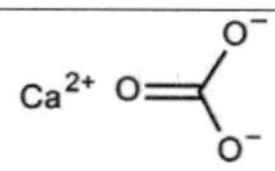	White nuances	[22]
Lapis lazuli	$(Na,Ca)_8(AlSiO_4)_6(S,SO_4,Cl)_{1–2}$	Blue nuances	[23]

Table 3.
Mineral source of natural dyes.

yellow ochre, raw sienna, malachite, ultramarine blue, azurite, gypsum, talc, and charcoal black serve as natural sources of dyes.

Table 3 delineates some prominent mineral sources of natural dyes. Limonite, commonly known as ochre, provides hues ranging from yellow to brown and red. Malachite, a vibrant green mineral composed of copper carbonate and hydroxide, yields striking green tones. Manganese, a metallic element, contributes to the creation of deep black shades. Cinnabar, characterized by its reddish hue and metallic luster, offers a spectrum of red shades. Azurite, a blue copper mineral often found alongside malachite, is achieving blue tones, while lead contributes to red nuances. Aragonite, typically colorless or white, is utilized for producing white shades. Lapis lazuli, a blue rock containing azurite, calcite, and pyrite, lends its distinctive blue hues to natural dyes [24].

2.4 Microbial and fugal origin

Various microorganisms exhibit the remarkable ability to produce pigments as secondary metabolites, contributing to the diverse palette of natural dyes. Bacteria such as Bacillus, Brevibacterium, Flavobacterium, Achromobacter, Pseudomonas, and Rhodococcus spp. are known for their pigment-producing capabilities. Notably, certain microorganisms have been identified for their production of indigo in response to petroleum products, underscoring their potential industrial applications [25].

Table 4 delineates some key microorganisms and their pigment-producing abilities. For instance, the red basidiomycetous yeasts *Agrobacterium aurantiacum* and *Xanthophyllomyces dendrorhous* are responsible for producing the orange–red pigment astaxanthin. Similarly, carotenoid canthaxanthin is synthesized by Bradyrhizobium strains and Halobacterium spp., offering a spectrum of vibrant hues [26]. Additionally, *Chromobacterium violaceum* produces violacein, a versatile pigment with applications in medicine, cosmetics, food, and textiles.

Furthermore, various microorganisms contribute to the production of essential compounds like beta-carotene and riboflavin. Phycomyces and Mucor circinelloides are utilized for beta-carotene synthesis, while microorganisms like *Serratia marcescens* and *Vibrio psychoerythrus* produce the pigment prodigiosin, known for its antibacterial, antimalarial, antineoplastic, and antibiotic properties [27]. Moreover, riboflavin

Microorganism	Pigments/ molecule	Molecular formula	Extract color
Agrobacterium aurantiacum	Astaxanthin	$C_{40}H_{52}O_4$	Pink to red
Bradyrhizobium Sp.	Canthaxanthin	$C_{40}H_{52}O_2$	Dark red
Flavobacterium Sp., Paracoccus zeazanthinifaciens	Zeaxanthin	$C_{40}H_{56}O_2$	Yellow
Corynebacterium insidiosum	Indigoidine	$C_{10}H_8N_4O_4$	Blue
Chormobacterium violaceum	Violacein	$C_{20}H_{13}N_3O_3$	Purple
Dunaliella salina	Β-carotene	$C_{40}H_{56}$	Red
Fusarium sporotrichioides	Lycopene	$C_{40}H_{56}$	Red
Haematococcus pluvialis	Astaxanthin	$C_{40}H_{52}O_4$	Red
Monascus purpureus	Monascin Ankaflavin	$C_{21}H_{26}O_5$	Red to yellow
Monascus roseus	Canthaxanthin	$C_{40}H_{52}O_2$	Orange to pink
Pacilomyces farinosus	Anthraquinone	$C_{14}H_8O_2$	Red
Phaffia rhodozyma	Astaxanthin	$C_{40}H_{52}O_4$	Pink to red
Rhodotorula sp. Rhodotorula glutinis	Torularhodin	$C_{40}H_{52}O_2$	Orange to red
Streptoverticillium rubrireticuli	Prodigiosin	$C_{20}H_{25}N_3O$	Red
Streptomyces echinoruber	Rubrolone	$C_{23}H_{23}NO_8$	Red

Table 4.
Microbial and fungal origin [26].

production by microorganisms such as Ascomycetes *Ashbya gossypii*, *Candida famata*, and *Bacillus subtilis* offers sustainable alternatives to traditional chemical synthesis methods, finding applications in a wide range of food and beverage products [28].

3. Extraction methods of natural dyes

The extraction process plays a pivotal role in isolating the desired color components from plants by breaking down their cell walls and facilitating their separation into the solvent medium [29]. In recent years, there has been a notable global shift toward eco-friendly and biodegradable materials. Within textile manufacturing and other industries such as cosmetics, medicine, and food, natural dyes and colors have emerged as favored alternatives to synthetic counterparts. However, extracting natural dyes presents a unique challenge due to the complex nature of plant matrices, which contain various non-dyed constituents alongside coloring materials [30]. Therefore, understanding the characteristics and solubility of these coloring materials is crucial before the extraction process. A variety of extraction methods are employed for natural coloring materials, including aqueous extraction, solvent extraction, enzymatic extraction, fermentation, alkaline or acid extraction, supercritical fluid extraction, as well as extraction utilizing ultrasonic energy or microwave technology. Each method offers distinct advantages and is selected based on factors such as the nature of the plant material and the desired outcome of the extraction process.

3.1 Aqueous extraction

The aqueous extraction method, a traditional approach for extracting colors from plants and other materials, involves soaking the colored materials in water, sometimes with the addition of salt, acid, alkali, or alcohol to enhance extraction efficiency [31]. Typically, the materials are first crushed into a powder form and left to steep overnight in a steel container to facilitate the breakdown of cell structures. Upon boiling, the compounds are carried away with steam due to their insolubility or slight solubility in water, with a vapor pressure of 100°C. Upon condensation, the extracted components are separated using an oil-water separator, followed by filtration using a trickling filter to isolate the dye from plant residues [32]. It is important to note that maintaining a low temperature during extraction is advisable, especially for temperature-sensitive dyes, as excessive heat can adversely affect color yield. The resulting dye can then be utilized in various textile applications.

3.2 Solvent extraction

Natural coloring materials can be extracted using various organic solvents such as acetone, petroleum ether, chloroform, ethanol, methanol, or their combinations like ethanol and methanol, water, and alcohol mixtures [30]. This method allows for the extraction of both water-soluble and water-insoluble substances from plant resources, resulting in higher extraction yields compared to aqueous methods. By adding acid or alkali to alcoholic solvents, hydrolysis of glycosides and subsequent release of coloring materials can be facilitated. Distillation enables easy removal and reuse of solvents, while conducting the extraction process at lower temperatures minimizes degradation risks. However, drawbacks include the presence of toxic residual solvents, contributing to greenhouse gas emissions, and the extract's poor solubility in water necessitating dyeing in an aqueous medium post-extraction. Additionally, co-extraction of waxy materials and chlorophylls can pose challenges [33].

3.3 Enzymatic extraction and fermentation

In recent years, there has been a notable surge in interest regarding the extraction of effective components from natural plants. Utilizing appropriate enzymes can facilitate the mild decomposition of plant tissues, leading to the rapid release of effective components and improved extraction rates. For instance, cellulase, known for degrading cellulose and hemicellulose, also modifies cell wall and cytoplasmic structures, thereby enhancing the diffusion of effective components into the extraction medium and improving pigment extraction efficiency. Enzyme activity is predominantly influenced by factors such as temperature and pH [33]. Compared to solvent extraction, enzymatic extraction offers advantages including milder extraction conditions and the preservation of the chemical and physical properties of active components. Enzymatic reactions can even alter the structure of certain compounds, such as Geniposide in natural gardenia yellow pigment, resulting in the production of gardenia red and blue pigments. Notably, enzymatic extraction demonstrates a 72% higher extraction rate for anthocyanins compared to solvent extraction [34]. This method proves effective for extracting dyes from the bark and roots of sturdy plants.

3.4 Alkaline or acid extraction

Natural dyes containing glycosides can be efficiently extracted using weak acids and alkalis, hastening the hydrolysis of glucosides and enhancing color yield [35]. Alkaline extraction is particularly effective for dyes with phenolic groups, yielding superior results [36]. Post-extraction, acids aid in precipitating lac dyes from lac pests and safflower leaves, while alkaline extraction stands as the optimal method for extracting lac dyes. However, the pH sensitivity of some natural dyes poses a drawback, potentially leading to a loss in color yield [35].

3.5 Supercritical fluid extraction

Supercritical fluid extraction offers a unique amalgamation of liquid and gas properties, boasting high density, viscosity, and solubility, along with lower surface tension. Despite its complexity, this mechanism effectively penetrates extraction materials' matrices, making it a valuable extraction tool [37]. Carbon dioxide (CO_2) stands out as a clean, safe, and abundant solvent alternative in supercritical fluid extraction due to its non-toxic nature and accessibility. With critical temperatures and pressures of 31.4°C and 1070 pounds per square inch (psi) or 73.8 bars, respectively, CO_2 supercritical extractions typically occur between 32°C and 49°C under pressures ranging from 1070 to 3500 psi. While CO_2 behaves akin to nonpolar organic solvents, the addition of cosolvents may aid in solubilizing slightly polar solutes. Notably, resulting extracts are light-colored and devoid of residual solvent traces and heavy metals, making them ideal for food and pharmaceutical purposes. However, drawbacks include high equipment costs and limitations in extracting polar substances [30].

3.6 Extraction with ultrasonic energy and microwave

Microwave and ultrasound-assisted extraction methods have revolutionized the extraction process by enhancing efficiency while reducing solvent usage, time, and temperature requirements. When plant materials containing natural dyes are subjected to ultrasound, the formation and collapse of small bubbles or cavitation occur within the solvent. This phenomenon generates extremely high temperatures and pressures, accelerating the extraction process by efficiently breaking down cell structures and releasing dye molecules. Additionally, the gentle nature of ultrasound allows for the extraction of heat-sensitive dye molecules at lower temperatures, preserving their integrity [26, 36]. Likewise, microwave-assisted extraction involves minimal solvent usage and the application of microwave energy to expedite the extraction process, yielding higher outputs in shorter durations. This innovative approach has garnered significant attention from researchers exploring new dye sources and optimizing extraction techniques [30].

4. Color fastness behavior of natural dyes

Color fastness refers to the resistance of a dyed material, such as fabrics or fibers, to maintain its color when subjected to various environmental factors like sunlight, washing, and other external agents. Achieving optimal color fastness with natural dyes involves considering several factors, including the type of dye used, the substrate (material being dyed), and the dyeing method employed. During discussions in this session, we briefly explored the diverse behaviors of color fastness. It is crucial to understand these factors to ensure that natural dyes retain their vibrancy and integrity over time, even when exposed to challenging conditions. By addressing these considerations, we can enhance the durability and longevity of dyed materials, providing both esthetic appeal and practical functionality. The color fastness of the dyed goods can be broadly categorized into (a) wet fastness and (b) light fastness (see **Figure 1**). However, some wet fastness properties are considered optional according to the application of the fabric. The wet fastness includes fastness to wash, water, perspiration, seawater, swimming pool, etc.

On the contrary, color fastness to light refers to the resistance against sunlight. This two fastness depend on the distinct factors. The wet fastness is the competition of dye molecules in a fiber and wash liquor or active agent as appropriate. It is the physio-chemical interaction aided by diffusion largely governed by the size of the dye molecules, amount of dye present in the fiber, and mode of interaction (covalent bond, salt linkages, polar bonds, etc.) between dye and fiber. In general, higher the depth of shade comparatively lower the wet fastness anticipated. The wet fastness can be improved by removing unfix dye molecules from the fiber surface or exaggerate the dye particle size by complex formation with different surfactants. On the other hand, light fastness is the resistance to fading administered by the photochemical reactions. It is the inherent attributes of dye chromogen and cannot be improved significantly with any additional treatments. In general, higher depth of shade experience the better light fastness as the photochemical fading is less pronounced inhibited by the large number dye molecules.

4.1 Color fastness to light and rubbing

The lightfastness of natural dyes is a critical consideration in textile dyeing, with most natural dyes exhibiting a lightfastness below BS grade 5, and the majority having a fastness rating below 4 [26]. Exposure to artificial light or daylight can lead to significant fading of natural dyes over time. Color fastness to rubbing, another important aspect of dye performance, varies from good to excellent for natural fibers [38]. The resilience of natural fibers to both rubbing and light exposure is vital for ensuring the longevity of dyed textiles. Several studies have investigated the color fastness properties of various natural dyes on different fabrics (see **Table 5**).

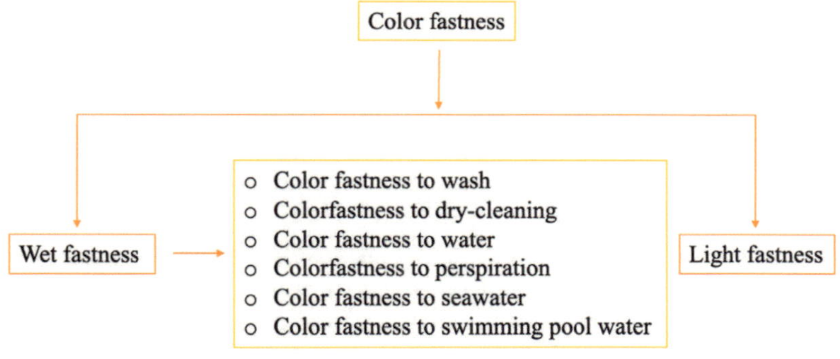

Figure 1.
Classification of color fastness for dyed fabric.

Dyes source	Fabric	Rubbing fastness		Light fastness	References
		Dry	Wet		
Henna	Cotton, wool	3–4	2–3	3–4	[39, 40]
Tulsi	Wool	4–5	4	5	[5]
Turmeric power	Silk, cotton	4	3–4	3	[41, 42]
Tamarind seed	Cotton, silk	4–5	3–4	—	[42, 43]
Amloki	Cotton	4	3–4	3–4	[44]
Mango leaves	Cotton, silk, polyester	4	3–4	3–4	[45]
Cochineal insects	Silk	4–5	4–5	4–5	[46]
Lac insect	Silk	4–5	4–5	4–5	[47]
Pomegranate	Cotton	3	2–3	3–4	[48]
Tea leaf	Cotton	3	2–3	3–4	[48]
Citrus aurantium	Lyocell	4	3–4	4	[49]
Kigelia africana	Linen	4–5	3–4	—	[50]
Triadica sebifera	Viscose	4	4–5	3–4	[51]
Thespesia populnea	Nylon	5	4–5	2	[52]

Table 5.
Color fastness to rubbing and light.

Alebeid and colleagues explored the use of henna dye on cotton and wool fabrics, noting that colorfastness to rubbing was better in dry conditions compared to wet conditions. Additionally, they observed good light fastness properties in all dyed fabrics [39]. Kamal Alebeid and Zhao utilized a different cationization method for cotton fabric and achieved a color fastness to light rating of 3–4 for henna extract dyed samples [40]. However, Owis achieved a higher rating of 4–5 when using henna dye on cotton and cotton-polyester blended fabrics with copper sulfate as a mordant [53]. Shah et al. investigated the application of natural dyes extracted from Tulsi on wool and reported improved color fastness to light and rubbing [5]. Yılmaz Şahinbaşkan focused on dyeing silk fabric with cochineal-derived natural dyes, employing alum as a pre-mordant and $FeSO_4$ as a post-mordant. They found better color fastness to both rubbing and light in pre-mordanted samples with alum, indicating the significance of the mordanting process in enhancing dye performance [46]. Mongkholrattanasit and his team explored the dyeing of silk fabric with lace insect dye using various mordants. They found that $CuSO_4$ yielded the best color and achieved good to very good color fastness to rubbing and moderate to good light fastness [47].

Studies on cotton fabrics dyed with tamarind seed extract and different mordants revealed good to excellent color fastness when dry and fair to good results when wet [43]. Similarly, Umar and colleagues achieved good color fastness to light when dyeing cotton and silk fabrics with tamarind fruit pod extract using alum, copper sulfate, and ferrous sulfates as mordants [54]. Islam and his research group experimented with natural dye sources such as amloki, haritaky, bahera, and arjuna on cotton fabric, observing better rubbing fastness but fair to moderate light fastness. They noted that the presence of tannic acid in these dyes contributed to improved fastness properties, especially in the absence of mordants [44]. Banna investigated mango leaf dye on silk, cotton, and polyester fabrics using various mordants, achieving excellent color fastness to rubbing and light on silk fabric but poorer results on cotton and polyester [45]. Islam dyed cotton fabric with pomegranate, henna, turmeric powder, and tea leaf extracts, observing the best colorfastness properties with henna dye but poor results with other dyes [55]. Hossain and his group demonstrated the dyeing of silk and jute fabric with the aqueous extract of coconut leaves and found the light fastness rating of 3–4 to 4–5 (very good) and rubbing fastness rating of 4–5 (excellent) [56].

Lyocell fabric dyed with orange peel extract showed excellent rubbing fastness and lightfastness, with ferrous sulfate demonstrating better performance compared to copper sulfate [49]. Linen fabric dyed with *kigelia africanna* flower extract using different mordants exhibited very good to excellent dry rubbing fastness and satisfactory levels of wet rubbing fastness [50]. Natural dye extracted from *Triadica sebifera* and applied on viscose fabric. $FeSO_4{\cdot}7H_2O$, $KAl(SO_4)_2{\cdot}12H_2O$, $SnCl_2{\cdot}2H_2O$, $CuSO_4{\cdot}5H_2O$, $ZnSO_4{\cdot}7H_2O$ are used as mordant. The values of dry and wet rubbing speed were found to be in the range of 4 and 5. Post-mordanting with Al^{3+} mordanting gave the results a light rating of 4, indicating fair lightfastness, while pre-mordanting and meta-mordanting gave results between 3 and 4. Amutha and his team dyed silk, cotton, and nylon fabrics with *Thespesia populnea* and *Terminalia arjuna* fruits, achieving excellent color fastness to rubbing with alum-mordanted samples but poor light fastness [52].

In conclusion, while natural dyes offer eco-friendly alternatives to synthetic dyes, their performance in terms of light and rubbing fastness varies depending on factors such as dye type, fabric substrate, and mordanting techniques. Further research and optimization of dyeing processes are necessary to enhance the overall durability of textiles dyed with natural dyes.

4.2 Color fastness to wash

Color fastness to wash is a crucial factor in evaluating the durability of dyed materials, measuring their resistance to washing. While natural dyes offer eco-friendly alternatives to synthetic ones, their wash fastness may not always match up. The efficacy of mordants and dyeing methods further influences wash fastness. Let us explore some key findings regarding the color fastness to wash of various natural fibers (see **Table 6**).

Owis conducted a comparative study involving onion skin and henna-extracted dyes on cotton, polyester, and cotton-polyester blends, with copper sulfate as the mordant. Results indicated higher color staining values for henna dye compared to onion skin dye. Additionally, natural dyes containing functional groups, such as lawson and anthocyanin, exhibited excellent wash fastness due to their ability to form covalent bonds with cotton fibers [53]. Furthermore, dye extracted from *Ocimum sanctum* (Tulsi) demonstrated very good wash fastness [40]. Bose and Nag achieved good color fastness to wash when dyeing cotton and silk fabrics with *Hibiscus rosa* [57]. Similarly, Khin reported an improvement in wash fastness when using turmeric powder as a natural dye source [41]. Aye & Kar observed fair fastness in washing with tamarind seed dye, with good staining on cotton and polyester. Post-mordanting proved effective in enhancing the wash fastness of cotton fabrics [43]. Dyeing cotton fabric with extracts from amloki, haritaky, bahera, and arjuna using ferrous sulfate and potash alum as mordants resulted in poor color change wash fastness but good staining, suggesting suitability for applications requiring minimal washing, such as wound bandages [45].

Dyes source	Fabric	Color fastness to wash	Reference
Henna	Cotton, polyester, and cotton/polyester blend	4–5	[53]
Hibiscus rosa	Cotton, silk	5	[5]
Turmeric power	Silk, cotton	4–5	[57]
Tamarind seed	Cotton, silk	3–4	[41, 42]
Amloki	Cotton	1–2	[43, 54]
Mango leaves	Cotton, silk, polyester	4–5	[44]
Cochineal insects	Silk	3–4	[45]
Lac insect	Silk	1–2	[47]
Pomegranate peel	Cotton	5	[8]
Citrus aurantium	Lyocell	4–5	[49]
Kigelia africana	Linen	4–5	[50]
Triadica sebifera	Viscose	4–5	[51]
Thespesia populnea	Nylon	4	[52]

Table 6.
Color fastness to wash.

Banna's research team found excellent wash fastness in silk fabric dyed with mango leaf extract but poor results for cotton and polyester fabrics, indicating variability in wash fastness depending on the fabric type and mordant used [45]. Burcu Yilmaz et al. achieved high wash fastness ratings of 4–5 when dyeing silk fabric with cochineal and gall oak [46]. However, eco-friendly dyeing of silk fabric with lac dye resulted in poor wash fastness ratings [47]. Dyeing cotton fabric with thyme and pomegranate peel using different mordants demonstrated varying wash fastness results. Pre-mordanted samples showed slightly lower wash fastness compared to samples without mordants [8]. Lyocell fabric dyed with orange peel using ferrous sulfate and copper sulfate as mordants showed varying washing fastness results depending on the concentration of the mordants. Ferrous sulfate exhibited better washing fastness compared to copper sulfate. Specifically, at a concentration of 4%, the washing fastness ratings were higher (4–5), whereas at a concentration of 2%, the ratings were slightly lower (3–4). Despite these variations, all samples experienced only a slight change in color, with ratings falling within the range of 4–5 [49].

In the case of linen fabric dyed with *Kigelia africana* flower, potash alum, copper sulfate, and ferrous sulfate were utilized as mordants. Interestingly, the washing fastness of the dyed fabrics remained consistently excellent across all samples, irrespective of the mordant used. There was no discernible difference in the ratings for any of the dyed samples. Furthermore, negligible dye removal and staining were observed, as reflected in the washing fastness ratings on the staining scale [51]. When tested with multi-fiber textiles such as acetate, cotton, nylon, polyester, acrylic, and wool, three different mordanting methods yielded identical colorfastness results. Both pre-mordanting and post-mordanting processes demonstrated excellent colorfastness to washing, particularly in polyester, acrylic, and wool fibers. Notably, the post-mordanting method resulted in the highest ratings of 4–5 for polyester, acrylic, and wool, indicating excellent wash fastness properties and slightly superior performance compared to the other two mordanting methods [52]. In their study, Amutha and their team focused on dyeing textiles, including cotton, silk, and nylon, using *Terminalia arjuna* and *Thespesia populnea* fruits. The color fastness to washing exhibited a good rating for silk and nylon fabrics. However, cotton fabric displayed poor color fastness to wash, suggesting variations in the performance of natural dyes across different types of textiles.

Overall, these findings underscore the importance of considering both the type of natural dye and the choice of mordant in achieving optimal wash fastness for dyed textiles. Further research into the development of sustainable dyeing processes and the optimization of mordanting techniques is warranted to enhance the wash fastness of natural dyes across various textile substrates.

4.3 Color fastness to perspiration

Color stability when exposed to perspiration is a crucial consideration for textiles in direct contact with human skin. Proper fixation of dyes to fibers can significantly enhance wash fastness and reduce bleeding or fading upon perspiration exposure. Various studies have explored the color fastness to perspiration of natural dyes on different types of fabrics, shedding light on their performance under such conditions. Some natural fiber's color fastness to perspiration is listed below (see **Table** 7).

Owis developed an economical dyeing process using henna and onion skin extracts on cotton, polyester, and cotton/polyester blended fabrics, employing copper sulfate as a mordant. Notably, henna dye exhibited superior color fastness to perspiration compared to onion skin dye, particularly at lower temperatures (>30°C), showcasing better color stability [53]. Vankar and Shukla conducted research on cotton and silk fabrics dyed with *Hibiscus rosa* extracts using various mordants. They found that cotton fabric treated with $CuSO_4$ mordant showed good color fastness to perspiration compared to wool fabrics, whereas $SnCl_2$-treated silk fabric demonstrated superior results [60].

Dyes source	Fabric	Perspiration		Reference
		Acid	Alkali	
Henna	Cotton, wool	4–5	2–5	[53]
Hibiscus rosa	Cotton, silk	4	3–4	[2]
Turmeric power	Silk, cotton	4	4–5	[41, 42]
Tamarind fruit pods	Cotton, silk	5	4–5	[54]
Mango leaves	Cotton, silk, Polyester	4	4–5	[45]
Cochineal insects	Silk	4–5	4–5	[46]
Lac insect	Silk	3–4	3–4	[47]
Pomegranate peel	Cotton	2	2	[58]
Citrus aurantium	Lyocell	4–5	4–5	[49]
Azadirachta indica	Linen	1–2	3–4	[5]
Triadica sebifera	Viscose	4	4	[51]
Onion outer shell	Nylon	4	2–3	[59]

Table 7.
Color fastness to perspiration.

Sarker and their team dyed silk fabric with turmeric powder using both natural and synthetic mordants. Interestingly, they obtained almost identical results for color fastness to perspiration regardless of the mordant used, indicating comparable performance between natural and synthetic options [42]. Umar and their research group focused on cotton and silk fabrics dyed with tamarind fruit pod extracts using different metal salts as mordants. They reported excellent to good color fastness to perspiration for alkaline extracts, with some variations observed depending on the mordant used [54]. Banna and their team explored the dyeing of cotton, silk, and polyester fabrics with mango leaf extracts using various mordants. While silk fabric exhibited excellent color fastness to perspiration across all mordanting options, cotton fabric showed slightly lower ratings [45]. Yilmaz and their team investigated the dyeing of silk fabric with natural dyes from cochineal insects and gall oak using alum and ferrous sulfate as mordants. They achieved good color fastness to perspiration, noting better results for acid perspiration compared to alkaline perspiration [46].

Mongkholrattanasit and their team dyed silk fabric with lac dye using different mordants and concentrations. While the results varied slightly, all treated samples exhibited fair to good color fastness to acid and alkaline perspiration [47]. Yang observed that fabrics dyed with natural mordants generally showed higher color staining fastness than color change fastness during washing and perspiration. Pomegranate peel dye, when mordanted with *Chaenomeles speciosa*, exhibited superior color staining fastness compared to color change fastness [58]. In acidic and alkaline solutions, lyocell fabric dyed with orange peel demonstrated good perspiration fastness, with minimal color changes observed when exposed to ferrous sulfate and copper sulfate [49]. Patel and their team utilized dyes from *Mangifera indica* and *Azadirachta indica* leaves on linen fabric, employing various natural mordants. The acid and alkaline perspiration fastness were reported to be fair to good [5]. Mia and their team dyed viscose fabric with natural dye extracted from *Triadica sebifera* using different mordants. Pre-mordanting and post-mordanting with Al^{3+} mordant yielded better results for acid and alkali perspiration compared to meta-mordanting [51]. Miah dyed nylon fabric with onion outer shell extract using different mordants. The color fastness to perspiration was notably good under acidic conditions, with no significant difference observed between the mordants used [59].

These studies collectively highlight the diverse performance of natural dyes and mordants concerning color fastness to perspiration, underscoring the importance of selecting appropriate dyeing techniques and mordants to ensure optimal performance in textile applications exposed to perspiration. Further research and development in this field are essential for enhancing the durability and sustainability of natural dyeing processes in the textile industry.

5. Challenges of natural dyes in terms of fastness

In comparison to synthetic dyes, natural dyes are sustainable and eco-friendly. However, they come with their own set of challenges, particularly in terms of fastness, which refers to the ability of a dye to retain its color when subjected to various external factors. As far as fastness is concerned, natural dyes face the following challenges.

5.1 Fastness properties

The performance of textiles is crucially determined by their colorfastness to light and washing, with additional importance placed on perspiration and rubbing, especially for clothing applications. While curtains necessitate excellent light fastness, a slightly lower washing fastness may suffice. However, natural dyes pose challenges due to their limited fastness properties, often falling short of modern textile requirements. The restrictions imposed by eco-standards on certain metal salts as mordants, such as chromium, copper, and tin, have further compounded the issue, narrowed the color range of natural dyes, and complicated the production of shades with adequate fastness qualities. Additionally, incorrect application procedures can lead to poor fastness properties, underscoring the need for optimized mordanting and dyeing techniques [30].

Natural dyes typically exhibit poor to medium light fastness due to chromophoric changes in their structure upon light absorption. Efforts to enhance light fastness have been explored extensively, aiming to mitigate the lack of chromophoric groups to dissipate resonance energy. Similarly, washing fastness is compromised by weak bonding between dyes and fibers, resulting in subpar resistance to detergent solutions. The influence of alkalinity on the color values of naturally dyed fabrics during washing has been studied, revealing alterations in hue and value under alkaline pH conditions. Despite these challenges, natural dyes generally demonstrate good to moderate rubbing fastness, as evidenced by studies evaluating various dye sources on cotton and jute fabrics. For instance, Samanta and colleagues reported favorable rubbing fastness for jackfruit wood, manjistha, red sandalwood, babool, and marigold dyes on cotton and jute substrates [61].

In summary, addressing the limitations of natural dyes necessitates a multifaceted approach involving improvements in mordanting and dyeing techniques, exploration of new dye sources, and continued research to enhance light and washing fastness properties. These endeavors are critical for advancing the viability and sustainability of natural dyeing practices in the textile industry [62].

5.2 Low yield

One significant obstacle is the low yield of natural dyes, which affects their efficiency and cost-effectiveness. Higher color yield is desirable as it enhances the longevity of the color and improves resistance to fading from light exposure, rubbing, and washing. Synthetic or inorganic mordants can be employed to increase color yield when [7]. Ali and Hussain observed variations in dye yield depending on the pH of the medium, with the maximum yield at pH 12 (alkaline) and the minimum at pH 3.4 (acidic) [63].

5.3 Color range limited

The color range offered by natural dyes is limited compared to synthetic counterparts. While red and yellow dyes have several sources, blue dyes are predominantly derived from indigo. The application processes of natural dyes differ, and their compatibility for mixing is restricted, thereby limiting the range of achievable colors. Indigo, for instance, requires a distinct application process, increasing both time and cost, even for secondary colors like green [30].

5.4 Color stability

Retaining the original color of natural dyes is another big challenge. As one of the first characteristics of a product, color is very important to consumers. In some cases, natural dyes have problems with color stability and brightness. Natural dyes are affected by factors such as light, temperature, and pH. So, it is hard to keep their original color. The color derived from anthocyanin is not stable when exposed to light and temperature [64].

5.5 Extraction process and time

The extraction process and time required for natural dyes are substantial compared to synthetic alternatives. Natural dyes often necessitate an additional mordanting step, making the dyeing process more time-consuming. While the use of raw dye-bearing materials ensures authenticity, it also prolongs the extraction process and requires specific setup, making it labor-intensive and incompatible with many commercial textile dyeing machines [30].

6. Preventive action for natural dyeing: improving fastness

Improving the fastness of natural dyes is crucial for enhancing the longevity and durability of dyed fabrics. Several preventive actions can be undertaken to achieve better color retention and resistance to fading.

6.1 Choice of right mordant

Mordants play a vital role in natural dyeing by facilitating the binding of dye molecules to fibers [65]. Experimenting with different mordants, such as alum, iron, and copper, allows for finding the optimal combination to improve color fastness. It is crucial to select the right mordant, as different mordants may affect the color outcome. Natural mordants, such as tamarind seed coat and *Emblica officinalis* tannins, have been shown to enhance the light fastness properties of dyes like turmeric, henna, madder, and pomegranate rind [37, 66]. Choosing the appropriate mordant can significantly enhance the fastness properties of natural dyes.

6.2 Enzyme treatment

Enzyme treatment offers an environmentally friendly alternative to conventional chemical processes in textiles. When applied to fabrics dyed with natural dyes, enzyme-treated fabrics exhibit improved colorfastness properties [67]. Studies have shown that pretreating wool with alkaline protease enzymes before dyeing resulted in significantly better color difference and color strength compared to untreated samples [68]. Similarly, enzyme treatments combined with ultrasound on cotton before dyeing with natural dyes have led to higher color efficiency and darker shades [69].

6.3 Co-pigmentation method

Co-pigmentation is a method used to enhance the durability and brightness of natural dyes. This technique involves combining two types of dyes, either in liquid–liquid or powder form. While the stability of color produced by co-pigmentation in powder form may be affected, dyes in liquid form tend to be brighter. However, the quality of color brightness in powder form can be improved by incorporating co-pigmentation additives and optimizing spray-drying processes [70].

6.4 Irradiation technology

Irradiation technologies, including ultrasound radiation, ultraviolet radiation, gamma radiation, electron beam irradiation, and plasma treatment, are gaining popularity in textile dyeing and finishing. These techniques have been shown to enhance dye uptake, fastness properties, and dyeability [71]. Ultraviolet radiation, for instance, affects only the chemical properties of the fabric's upper layers without altering its bulk properties. By increasing dye uptake and improving fabric fastness, ultraviolet treatments contribute to producing deeper shades while maintaining fabric integrity [72].

7. Discussions

Natural dyes offer a plethora of rich and intricate hues, embodying centuries of tradition and cultural heritage. While their initial cost may appear steep, their potency and concentration render them remarkably economical. As interest in traditional dyes persists, the quest for novel sources intensifies, driving a renaissance in natural dye research and innovation [2]. Scientists and researchers worldwide are diligently enhancing natural dyes, marking an era of revitalization for this ancient craft. The allure of natural dyes extends beyond their vibrant colors, encompassing environmental sustainability, biodegradability, and safety for human health. Unlike their synthetic counterparts, natural dyes pose no threat of allergies or carcinogenicity, aligning with the ethos of green consumerism. Recognizing their pivotal role in fostering sustainable development, societies increasingly embrace natural dyes and their derivatives as indispensable elements of eco-friendly living.

Despite their undeniable benefits, the utilization of natural dyes remains modest, constituting a mere 1% of the textile dyeing market, primarily driven by artisans and small-scale entrepreneurs. While manufacturers across the globe leverage online platforms to market their wares, the trade in natural dye-bearing materials and extracts remains niche [30]. Urgent research imperatives revolve around enhancing color stability and vibrancy, addressing consumer preferences for commercial products. Innovations abound in natural dye stabilization processes, exemplified by Arabi gum's role in preserving anthocyanin pigments in beverages. Through hydrogen bonding with anthocyanins and glycoprotein fractions, Arabi gum fortifies color stability, while caffeic acid enhances color intensity by facilitating anthocyanin solubility [73]. Moreover, advancements in extraction methodologies promise cost-effective and eco-friendly solutions, circumventing the need for toxic chemicals prevalent in synthetic dye production.

The allure of natural dyes is further amplified by their reduced environmental footprint, as both processing and disposal entail minimal ecological impact [33]. Modern extraction techniques and user-friendly applications are poised to attract diverse industries to embrace natural dyes, heralding a future where sustainability and esthetics harmonize seamlessly. Through concerted research endeavors, the potential of natural dyes to transform industries while safeguarding the environment becomes increasingly palpable.

8. Conclusions

In conclusion, the resurgence of natural dyes stands as a beacon of hope in the realm of textile coloring, offering a sustainable alternative to their synthetic counterparts. Unlike non-biodegradable and hazardous azo dyes, natural dyes boast biodegradability and pose minimal risk to human health and the environment. Their versatility is unmatched, enabling the creation of a myriad of color shades through varied mordants and processes, with some dyes even deepening in hue over time.

The popularity of natural dyes is further bolstered by their soft, lustrous shades and skin-friendly, non-toxic properties. Moreover, the labor-intensive process of cultivating, extracting, and applying natural dyes not only ensures superior quality but also generates employment opportunities, contributing to socioeconomic sustainability. By embracing natural dyes, we not only safeguard the environment from the harmful effects of synthetic dyes but also uphold centuries-old traditions of craftsmanship and innovation. Through meticulous research and continuous improvement efforts, we endeavor to unlock the full potential of natural dyes, ensuring their enduring legacy in textile dyeing. This chapter has endeavored to shed light on the diverse sources of natural dyes, their fastness properties, and the ongoing endeavors to enhance their durability, paving the way for a more vibrant and sustainable future in the textile industry.

Acknowledgement

The authors acknowledge that AI (Grammarly) has helped to improve the language to some extent.

Funding

None.

Conflict of interest

The authors declare no conflict of interest.

Author details

Tarikul Islam[1,2]*, Kazi Md. Rashedul Islam[1], Shahin Hossain[3], M. Abdul Jalil[4]*
and M. Mahbubul Bashar[5]

1 Department of Textile Engineering, Jashore University of Science and Technology, Jashore, Bangladesh

2 Department of Textiles, Merchandising, and Interiors, University of Georgia, Athens, Georgia, United States

3 Department of Environmental Sciences, BGMEA University of Fashion and Technology, Dhaka, Bangladesh

4 Department of Natural Sciences, BGMEA University of Fashion and Technology, Dhaka, Bangladesh

5 Department of Textile Engineering, Mawlana Bhashani Science and Technology University, Tangail, Bangladesh

*Address all correspondence to: mti@just.edu.bd; tarikul@uga.edu
and abduljalil@buft.edu.bd

References

[1] Hossain S, Jalil MA, Kader A, et al. Excellent dyeing properties of a natural dye extracted from the leaves of *Phoenix dactylifera* Linn on cotton and silk fabrics. Journal of Natural Fibers. 2022;**19**:9803-9812

[2] Islam SU, Majumdar A, Butola BS, editors. Advances in Healthcare and Protective Textiles. USA: Elsevier; 2023

[3] Tamilarasi A. Classification and types of natural dyes: A brief review. International Journal of Creative Research Thoughts. 2021;**9**:527-532

[4] ur Rehman F, Adeel S, Qaiser S, et al. Dyeing behaviour of gamma irradiated cotton fabric using Lawson dye extracted from henna leaves (*Lawsonia inermis*). Radiation Physics and Chemistry. 2012;**81**:1752-1756

[5] Shah SR, Patel BH. Characterization and fixation of *Ocimum sanctum* extract on wool fabric. Bangladesh Textile Today. 2013;**6**:31-34

[6] Arnone A, Camarda L, Merlini L, et al. Colouring matters of the West African red woods *Pterocarpus osun* and *P. soyauxii*. Structures of santarubins A and B. Journal of the Chemical Society, Perkin Transactions 1. 1977;**1977**:2116

[7] Gupta VK. Fundamentals of natural dyes and its application on textile substrates. In: Chemistry and Technology of Natural and Synthetic Dyes and Pigments. London, UK: IntechOpen; 2019

[8] Davulcu A, Benli H, Şen Y, et al. Dyeing of cotton with thyme and pomegranate peel. Cellulose. 2014;**21**:4671-4680

[9] Menegazzo MAB, Giacomini F, de Barros MASD. Study of wool dyeing with natural dye extracted from chamomile flowers. Journal of Natural Fibers. 2020;**17**:271-283

[10] Yusuf M, Khan SA, Shabbir M, et al. Developing a shade range on wool by madder (*Rubia cordifolia*) root extract with gallnut (*Quercus infectoria*) as biomordant. Journal of Natural Fibers. 2016;**14**:1-11

[11] Harbourne N, Jacquier JC, O'Riordan D. Optimisation of the extraction and processing conditions of chamomile (*Matricaria chamomilla* L.) for incorporation into a beverage. Food Chemistry. 2009;**115**:15-19

[12] Galappaththi M, Patabendige N. Cochineal chemistry, related applications and problems: A mini review. Academia Letters. Epub ahead of print 4 February 2022. DOI: 10.20935/AL1792

[13] Kamel MM, El-Shishtawy RM, Yussef BM, et al. Ultrasonic assisted dyeing III. Dyeing of wool with lac as a natural dye. Dyes and Pigments. 2005;**65**:103-110

[14] Ziderman II. The biblical dye tekhelet and its use in Jewish textiles. Dyes in History and Archaeology. 2008;**21**:36-44

[15] Lech K, Jarosz M. Identification of polish cochineal (*Porphyrophora polonica* L.) in historical textiles by high-performance liquid chromatography coupled with spectrophotometric and tandem mass spectrometric detection. Analytical and Bioanalytical Chemistry. 2016;**408**:3349-3358

[16] Tamilarasan MG, Ramasubramaniam M. Study on extraction of cephalopod's ink for dyeing of textile. Global Journal of Engineering

Science and Research Management. 2019;**6**:31-37

[17] Chungkrang L, Bhuyan S. Natural dye sources and its applications in textiles: A brief review. International Journal of Current Microbiology and Applied Sciences. 2020;**9**:261-269

[18] Pargai D, Jahan S, Gahlot M. Functional properties of natural dyed textiles. In: Chemistry and Technology of Natural and Synthetic Dyes and Pigments. London, UK: IntechOpen; 2020. Epub ahead of print 30 September 2020. DOI: 10.5772/intechopen.88933

[19] Heinfling A, Martínez MJ, Martínez AT, et al. Transformation of industrial dyes by manganese peroxidases from Bjerkandera adusta and Pleurotus eryngii in a manganese-independent reaction. Applied and Environmental Microbiology. 1998;**64**:2788-2793

[20] Nöller R. Cinnabar reviewed: Characterization of the red pigment and its reactions. Studies in Conservation. 2015;**60**:79-87

[21] Vetter W, Latini I, Schreiner M. Azurite in medieval illuminated manuscripts: A reflection-FTIR study concerning the characterization of binding media. Heritage Science. 2019;7:1-10

[22] de Assis Filho RB, Baptisttella AMS, de Araujo CMB, et al. Removal of textile dyes by benefited marine shells wastes: From circular economy to multi-phenomenological modeling. Journal of Environmental Management. 2021;**296**:113222

[23] Emerson D. Lapis lazuli – The most beautiful rock in the world. Preview. 2015;**2015**:63-73

[24] Singh K, Kumar P, Singh NV. Natural dyes: An emerging ecofriendly solution for textile industries. Pollution Research. 2020;**39**:87-94

[25] Babitha S. Microbial pigments. In: Biotechnology for Agro-Industrial Residues Utilisation. Dordrecht: Springer Netherlands; 2009. pp. 147-162

[26] Malik K, Tokkas J, Goyal S. Microbial pigments: A review. International Journal of Microbial Resource Technology. 2012;**1**:361-365

[27] Khanafari A, Assadi MM, Fakhr FA. Review of prodigiosin, pigmentation in *Serratia marcescens*. Online Journal of Biological Sciences. 2006;**6**(1):1-13

[28] Stahmann K-P, Revuelta JL, Seulberger H. Three biotechnical processes using *Ashbya gossypii*, *Candida famata*, or *Bacillus subtilis* compete with chemical riboflavin production. Applied Microbiology and Biotechnology. 2000;**53**:509-516

[29] Habib N, Akram W, Adeel S, et al. Environmental-friendly extraction of Peepal (Ficus Religiosa) bark-based reddish brown tannin natural dye for silk coloration. Environmental Science and Pollution Research International. 2022;**29**:35048-35060

[30] Raja SS, M. AS. Natural dyes: Sources, chemistry, application and sustainability issues. Textile Science and Clothing Technology. 2014;**1**:37-80

[31] Safapour S, Sadeghi-Kiakhani M, Eshaghloo-Galugahi S. Extraction, dyeing, and antibacterial properties of *Crataegus elbursensis* fruit natural dye on wool yarn. Fibers and Polymers. 2018;**19**:1428-1434

[32] Lokesh P, Swamy MK. Extraction of natural dyes from *Spathodea*

campanulata and its application on silk fabrics and cotton. Der Chemica Sinica. 2013;**4**:111-115

[33] Salauddin SM, Mia R, Haque MA, et al. Review on extraction and application of natural dyes. Textile & Leather Review. 2021;**6257**:1-16

[34] Shirsath SR, Sonawane SH, Gogate PR. Intensification of extraction of natural products using ultrasonic irradiations—A review of current status. Chemical Engineering and Processing: Process Intensification. 2012;**53**:10-23

[35] Yusuf M, Shabbir M, Mohammad F. Natural colorants: Historical, processing and sustainable prospects. Natural Products and Bioprospecting. 2017;**7**:123-145

[36] Wang L. Study on extracting natural plant dyestuff by enzyme-ultrasonic method and its dyeing ability. Journal of Fiber Bioengineering and Informatics. 2009;**2**:25-30

[37] Prabhu KH, Bhute AS. Plant based natural dyes and mordants: A review. Journal of Natural Product and Plant Resources. 2012;**2**:649-664

[38] Padfield T, Landi S. The light-fastness of the natural dyes. Studies in Conservation. 1966;**11**:181

[39] Alebeid OK, Zhao T, Seedahmed AI. Dyeing and functional finishing of cotton fabric using henna extract and TiO_2 nano-sol. Fibers and Polymers. 2015;**16**:1303-1311

[40] Alebeid KO, Zhao T. An eco friendly dyeing of cotton fabric using henna extract. International Journal on Advanced Science, Engineering and Information Technology. 2017;**5**:2321-9009

[41] Khin OO, Oo ', Yee S. A study on the fastness properties of cotton fabric dyed with turmeric dyestuff. International Journal For Innovative Research In Multidisciplinary Field. 2017;**3**:135-138

[42] Sarker P, Hosne Asif AKMA, Rahman M, et al. Green dyeing of silk fabric with turmeric powder using tamarind seed coat as mordant. Journal of Materials Science and Chemical Engineering. 2020;**08**:65-80

[43] Mar AA, Thu AKCM. Study on the extraction and utilization of natural dye from tamarind seed testa [doctoral dissertation]. MERAL Portal. Mayanmar: University of Mandalay

[44] Islam T, Ahmed M, Karim MR, et al. Assessment of fastness properties of knitted cotton fabric dyed with natural dyes: A sustainable approach of textile coloration. Journal of Textile Engineering & Fashion Technology. 2019;**5**:177-182

[45] Uddin Banna B, Mia R, Sharmin Tanni K, et al. Effectiveness of dyeing with dye extracted from mango leaves on different fabrics by using various mordants introduction. North American Academic Research. 2019;**2**:123-143

[46] Yılmaz Şahinbaşkan B, Karadag R, Torgan E. Dyeing of silk fabric with natural dyes extracted from cochineal (Dactylopius coccus Costa) and gall oak (Quercus infectoria Olivier). Journal of Natural Fibers. 2018;**15**:559-574

[47] Mongkholrattanasit R, Saiwan C, Rungruangkitkrai N, et al. Ecological dyeing of silk fabric with lac dye by using padding techniques. Journal of the Textile Institute. 2015;**106**:1106-1114

[48] Islam MS, Alam SMM, Ahmed S. Attaining optimum values of colourfastness properties of sustainable dyes on cotton fabrics. Fibres and Textiles in Eastern Europe. 2020;**28**:110-117

[49] Tayyab N, Sayed RY, Faisal R, et al. Dyeing and colour fastness of natural dye from *Citrus aurantium* on lyocell fabric. Industria Textila. 2020;**71**:350-356

[50] Sheikh J, Singh N, Pinjari D. Sustainable functional coloration of linen fabric using *Kigelia Africana* flower colorant. Journal of Natural Fibers. 2021;**18**:888-897

[51] Mia R, Minhajul Islam M, Ahmed T, et al. Natural dye extracted from *Triadica sebifera* in aqueous medium for sustainable dyeing and functionalizing of viscose fabric. Clean Engineering and Technology. 2022;**8**:100471

[52] Amutha K, Grace Annapoorani S, Sudhapriya N. Dyeing of textiles with natural dyes extracted from *Terminalia arjuna* and *Thespesia populnea* fruits. Industrial Crops and Products. 2020;**148**:112303

[53] Owis AS. An economical dyeing process for cotton, polyester and cotton/polyester blended fabrics. Journal of Textile and Apparel, Technology and Management. 2010;**6**:1-11

[54] Umar IA, Bn Muh'd nor MN, Wong YC. Fastness properties of colorant extracted from locust beans fruits pods to dye cotton and silk fabrics. Environmental Management and Sustainable Development. 2013;**2**:67-73. DOI: 10.5296/emsd.v2i1.3228

[55] Shariful Islam SM, Alam M, Akter S. Investigation of the color fastness properties of natural dyes on cotton fabrics. Fibers and Textiles. 2020;**27**:1-6

[56] Hossain S, Jalil MA, Mahmud RU, et al. Natural dyeing of silk and jute fabric with the aqueous extract of coconut leaves – An eco-friendly approach. Pigment & Resin Technology. 2023. Epub ahead of print March 2023. DOI: 10.1108/PRT-10-2022-0125

[57] Bose S, Nag S. Isolation of natural dyes from the flower of *Hibiscus rosa-sinensis*. American Journal of PharmTech Research. 2012;**2**:761-770

[58] Yang R, Li J, Cheng G, et al. Textiles dyeing with pomegranate (*Punica granatum*) Peel extract using natural mordant. Journal of Natural Fibers. 2023;**20**:1-12. DOI: 10.1080/15440478.2023.2282056

[59] Miah MR, Zakaria HMA, et al. Eco-dyeing of nylon fabric using natural dyes extracted from onion outer shells: Assessment of the effect of different mordant on color and fastness properties. International Journal of Scientific and Engineering Research. 2016;**7**:1030-1043

[60] Vankar PSSJ. Natural dyeing with dark pink flowers of *Quisqualis indica*. Asian Dyers. 2011;**58**:61-63

[61] Samanta A, Agarwal P, Datta S. Dyeing of jute and cotton fabrics using jackfruit wood extract: Part I—Effects of mordanting and dyeing process variables on colour yield and colour fastness properties. Indian Journal of Fibre & Textile Research. 2007;**32**:466-476

[62] Cook CC. Aftertreatments for improving the fastness of dyes on textile fibres. Review of Progress in Coloration and Related Topics. 1982;**12**:73-89

[63] Ali S, Hussain T, Nawaz R. Optimization of alkaline extraction of natural dye from henna leaves and its dyeing on cotton by exhaust method. Journal of Cleaner Production. 2009;**17**:61-66

[64] Qian B-J, Liu J-H, Zhao S-J, et al. The effects of gallic/ferulic/caffeic acids on colour intensification and anthocyanin stability. Food Chemistry. 2017;**228**:526-532

[65] Hossain S, Jalil MA, Islam T, et al. Enhancement of antibacterial and UV protection properties of blended wool/acrylic and silk fabrics by dyeing with the extract of *Mimusops elengi* leaves and metal salts. Heliyon. 2024;**10**:e25273

[66] Zia KM, Adeel S, Rehman F, et al. Influence of ultrasonic radiation on extraction and green dyeing of mordanted cotton using neem bark extract. Journal of Industrial and Engineering Chemistry. 2019;**77**:317-322

[67] Madhu A, Chakraborty JN. Developments in application of enzymes for textile processing. Journal of Cleaner Production. 2017;**145**:114-133

[68] Raja ASM, Thilagavat G. Influence of enzyme and mordant treatments on the antimicrobial efficacy of natural dyes on wool materials. Asian Journal of Textile. 2011;**1**:138-144

[69] Benli H, Bahtiyari Mİ. Use of ultrasound in biopreparation and natural dyeing of cotton fabric in a single bath. Cellulose. 2015;**22**:867-877

[70] Harsito C, Prabowo AR, Prasetyo SD, et al. Enhancement stability and color fastness of natural dye: A review. Open Engineering. 2021;**11**:548-555

[71] Pizzicato B, Pacifico S, Cayuela D, et al. Advancements in sustainable natural dyes for textile applications: A review. Molecules. 2023;**28**:5954

[72] Shahid-ul-Islam MF. High-energy radiation induced sustainable coloration and functional finishing of textile materials. Industrial and Engineering Chemistry Research. 2015;**54**:3727-3745

[73] Brouillard R, Dangles O. Anthocyanin molecular interactions: The first step in the formation of new pigments during wine aging? Food Chemistry. 1994;**51**:365-371

Chapter 3

Food-Grade Dye Extracts/ Photosensitisers from Non-staple Plants: Prospects for Eco-friendly Low-Cost Food Disinfection and Topical Biomedical Applications

Hussaini Majiya

Abstract

The used of food-grade photosensitisers for controlling microorganisms promised to be a viable and sustainable alternative strategy to the conventional methods which are complex and operationally, chemically and energy intensive, thereby making them expensive. Also, the quest/preference by food consumers for green technology especially in food disinfection and sanitation is making use of food-grade photosensitisers as suitable alternative for the control of microorganisms. Food-grade photosensitisers/dye extract from non-staple plants could be used for photodynamic sanitation of foods and contact surfaces and treatment of localised topical wound and skin diseases. There are many types of food-grade photosensitisers and dye extracts that can be sourced abundantly from non-staple plant extracts that have got long history of safe use as medicines and foods by humans. These photosensitisers of plant origin are diverse groups of compounds including curcumoids, alkaloids, anthraquinones, perylenequinones, flavins, chlorins, thiophenes, carotenoids, terpenes, and anthocyanins. This chapter discusses basics and targets of photodynamic inactivation in microorganisms. It also discussed and outlined food-grade photosensitisers from plant sources and their photodynamic antimicrobial activity. The chapter concludes by discussing the new thinking/perspectives, challenges and prospects of using food-grade dye extracts for disinfection of foods and contact surfaces.

Keywords: food-grade photosensitisers, photodynamic inactivation, non-staple plants, reactive oxygen species, singlet oxygen, food disinfection

1. Introduction

Microorganisms in food and contact surfaces are controlled to prevent food spoilage and transmission of diseases [1–4]. The conventional means of controlling spoilage microorganisms and pathogens can be broadly grouped into thermal and non-thermal means. The thermal strategies will include boiling, pasteurisation, tyndallisation, sun drying and the use of low temperature (refrigeration). The non-thermal strategies include the use of chemicals, antibiotics, irradiation, ultra-high voltage pulse electric field, high hydrostatic pressure and high-intensity ultrasound and membrane filtration. These conventional and emerging microbial control methods are effective, but they are operationally complex, chemically, and energy-intensive, and therefore, ultimately costly [1, 3–5]. Thermal control of microorganisms in foods can lead to loss of original foods nutrients and flavour due to elevated temperatures [5]. Antibiotics are overused in food and biomedical applications; this has been implicated in the development of bacterial resistance and therefore it is being phased out, especially in food disinfection and sanitation [5–8]. The irradiation and pulse electric field means of controlling microorganisms are usually limited due to expensive equipment and large energy consumption [5]. The high-intensity ultrasound which is one of the emerging non-thermal microbial control strategies can generate free radicals, denature proteins, lead to hotspots and release metal ions [5]. The higher doses of ionising radiation may cause slight changes in the colour of beef, pork, and poultry meat [5].

Aside from numerous drawbacks of the conventional microbial control methods (especially use of chemicals and radiations), food consumers the world over are tending towards preferring clean and green technologies as alternatives to synthetic chemicals in the preservation and disinfection of foods and contact surfaces [3, 4, 7]. Also, although most of the foods/crops/plants are cultivated and produced in the rural areas of developing countries, the conventional microbial control methods are not usually sustainably available for use because of lack of power infrastructures or erratic power supply – this usually lead to huge annual post-harvest losses of the produce due to microbial spoilage and deterioration [3]. One of the microbial control alternatives proposed, especially for places with abundant free sunlight, is the use of food-grade photosensitisers for photodynamic disinfection of surfaces and foods including fresh produce [2–9]. Food-grade photosensitisers from non-staple plants can be used in photosensitisation process, which is one of the alternative clean and green technologies preferred by consumers to conventional food sanitising synthetic chemicals.

This chapter covers and discusses the prospects of using food-grade photosensitisers from non-staple plants for eco-friendly and low-cost food disinfection and topical biomedical applications. The chapter starts by highlighting the photodynamic effect and photodynamic inactivation (PDI). It then highlights the characteristics of major photosensitisers used in the photodynamic inactivation of microorganisms. It then discusses the biological targets of singlet oxygen and other reactive oxygen species in microorganisms. It then discusses food-grade photosensitisers based on their plant sources, availability, traditional usage, photodynamic and non-photodynamic antimicrobial characteristics and suitability for disinfection/sanitation of foods including fresh produce (fruits and vegetables) and topical biomedical applications. The chapter also then emphasises the need to test and know the photostability and photodynamic antimicrobial characteristics/performance of the dye extracts/photosensitisers (PS) in different solutions/buffers/diluents to determine their suitability for environmental and or biological applications. The chapter concludes by discussing the new thinking/ perspectives, challenges and prospects of using food-grade dye extracts for disinfection of foods and contact surfaces.

2. Photodynamic effect and photodynamic inactivation

Photodynamic effect is a resultant generation of reactive oxygen species (ROS) (singlet oxygen, superoxide, peroxide, and hydroxyl radical/ion) from the interaction of three things which are photosensitiser, visible light and molecular oxygen (**Figure 1**) [10, 11]. Photodynamic inactivation is a process in which the photodynamic effect generated ROS oxidise and cause irreversible damage to cellular biomolecules including proteins, lipids, nucleic acid and other components which ultimately lead to inactivation of microorganisms (**Figure 1**) [10, 11]. Therefore, photodynamic effect is a precursor and prerequisite of photodynamic therapy and inactivation.

Photodynamic effect is an indirect photochemical reactions (photosensitised reactions) in which the excited triplet state photosensitiser molecule transfer its energy to oxygen either through type 1 photoreaction (charge transfer) or type 2 photoreaction (electronic energy transfer) (**Figure 1**) [10, 11]. Type 1 photoreaction lead to the generation of superoxide, peroxide, and hydroxyl radical/ion while type 2 photoreaction lead to the generation of singlet oxygen. These two transfer mechanisms are non-radiative, which prevents the emission of radiation by the excited photosensitiser molecule, and this is called quenching [10, 11]. The oxygen, one of the quenchers, becomes excited and underwent various photophysical and photochemical processes based on its properties [10, 11]. The energy transfer occurs before the excited triplet state photosensitiser molecule can radiate fluorescence, and the acceptor molecule (oxygen) is thus excited indirectly, thereby undergoing various photophysical and photochemical processes called photosensitisation or photosensitised reactions (**Figure 1**) [10, 11]. The typical feature of a photosensitised reaction is that the photosensitiser remains unchanged while the oxygen undergoes chemical reactions [10, 11].

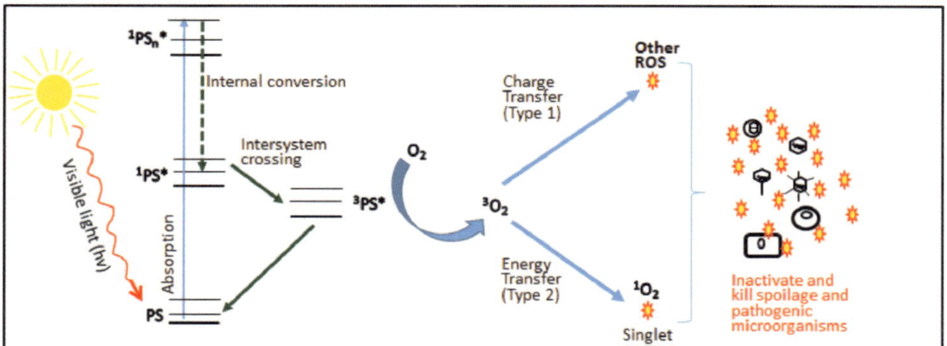

PS = photosensitiser

Figure 1.
A Jablonski energy scheme showing the interaction of light, photosensitiser molecule and molecular oxygen in a photosensitised reaction. Photosensitiser is the ground state photosensitiser molecule; $^1PS^*$ is the singlet state photosensitiser molecule; $^3PS^*$ is the triplet state photosensitiser molecule; O_2 is the molecular oxygen; $^3O_2$ is the triplet oxygen; $^1O_2^*$ is the singlet oxygen; other ROS: O_2. is the superoxide anion; $O_2.^{-2}$ is the peroxide; .OH is the hydroxyl radical; OH- is the hydroxyl ion [10].

Oxygen has a paramagnetic property (due to the presence of unpaired electrons) and is responsible for its efficient quenching ability. The molecule of a photosensitiser at the ground state absorbs visible light and gets excited to singlet state and then to triplet state where it interacts and gets quenched by the oxygen, which permeates most organic matter in liquid/solution, semi-solid and solid surfaces. This quenching results in the formation of ROS, which acts as an oxidising agent for the organic molecules present in the solution [10, 11]. This is the basis for photodynamic inactivation (PDI) of spoilage and pathogenic microorganisms and photodynamic Therapy (PDT) of cancers, dental, skin and autoimmune diseases.

3. Photosensitisers previously used for the photodynamic inactivation of microorganisms

Photosensitisers are generally dyes/coloured compounds/molecules with the ability to absorb light to get excited to form a long-lived excited triplet state. Generally, most photosensitisers identified and used for photodynamic inactivation of microorganisms were mainly porphyrines, phthalocyanins, Rose Bengal, methylene blue, curcumin, chlorophyllin, hypericin, and anthocyanins [10, 11]. For a dye/coloured compound to serve as a good photosensitiser, it has to exhibit some desired properties including having good solubility in water, intense absorption of light in the visible region extending to the near infra-red, having good stability upon prolonged storage in aqueous solution, with few or no side photoreactions, having high triplet yield and efficient production of separate ion products upon irradiation in the presence of oxygen [10, 11].

Rose Bengal, eosin and methylene are organic dyes which are good photosensitisers that that can get excited to triplet states of appropriate energies for oxygen sensitisation. These dyes exhibit intense absorption within the visible spectrum and show high singlet oxygen yields [10, 11]. The same is true for porphyrins and phthalocyanins because they have almost all the desired properties of a good photosensitiser. Also, since porphyrins and phthalocyanins can be present in natural systems, it is believed that the dyes might generally lack cytotoxicity in the dark which is good in some applications where only photosensitisation is required [10, 11]. Curcumin, chlorophyllin, hypericin, and anthocyanins are examples of food-grade photosensitisers and are thought to be most suitable for use in the PDI of spoilage and pathogenic microorganisms in food disinfection/sanitation and preservation [3].

Most microorganisms have net negative charge and therefore, cationic photosensitisers are more efficient in photodynamic inactivation of bacteria and viruses. This is because positive charge of a photosensitiser favours its binding to critical cellular sites and, with appropriate exposure to light, ultimately leads to loss of cell viability. Also, the strong attraction between the positively charged photosensitiser and negatively charged microorganism increases the proximity of the microorganism to the singlet oxygen generated. Therefore, it is generally expected that rate and extent of photodynamic inactivation is faster and higher with cationic photosensitisers compared to anionic and neutral photosensitisers [10, 12, 13].

4. Biological targets of singlet oxygen and other ROS

In order to understand the mechanisms and limitations of photodynamic inactivation of microorganisms, it is important to note that ROS have a short life and high reactivity, which means they can cause damage only to the surrounding molecules close to the points of their generation [10, 14, 15]. Singlet oxygen is reported to be the ROS mostly responsible for the photodynamic inactivation of bacteria and viruses, as it can react with biomolecules [10]. The singlet oxygen reacts rapidly with proteins with a bimolecular rate constant ranging from 105 to 109 $M^{-1}\ S^{-1}$ [10, 15–17]. This singlet oxygen-proteins bimolecular rate constant is higher than singlet oxygen with other biomolecules. The singlet oxygen-RNA reaction rate constant ranges from 104 to 106 $M^{-1}\ S^{-1}$ [10, 15–17]. The interaction of singlet oxygen with potential targets can either be by physical quenching or chemical modification, which usually results in irreversible changes [10, 15–17]. The physical quenching is only observed in tryptophan while the chemical modification is observed in almost all amino acids [10, 15–17]. The bimolecular rate constant and oxidation effects and products vary among different amino acids [10, 15–17]. The Amino acids tryptophan, histidine, methionine, cysteine and tyrosine are the most susceptible to singlet oxygen-mediated oxidation [10, 15–17]. The targets of photodynamic inactivation in microorganisms are most likely the side chains of some α-amino acids (mainly the aromatic and sulphur-containing amino acid residues), unsaturated lipids and

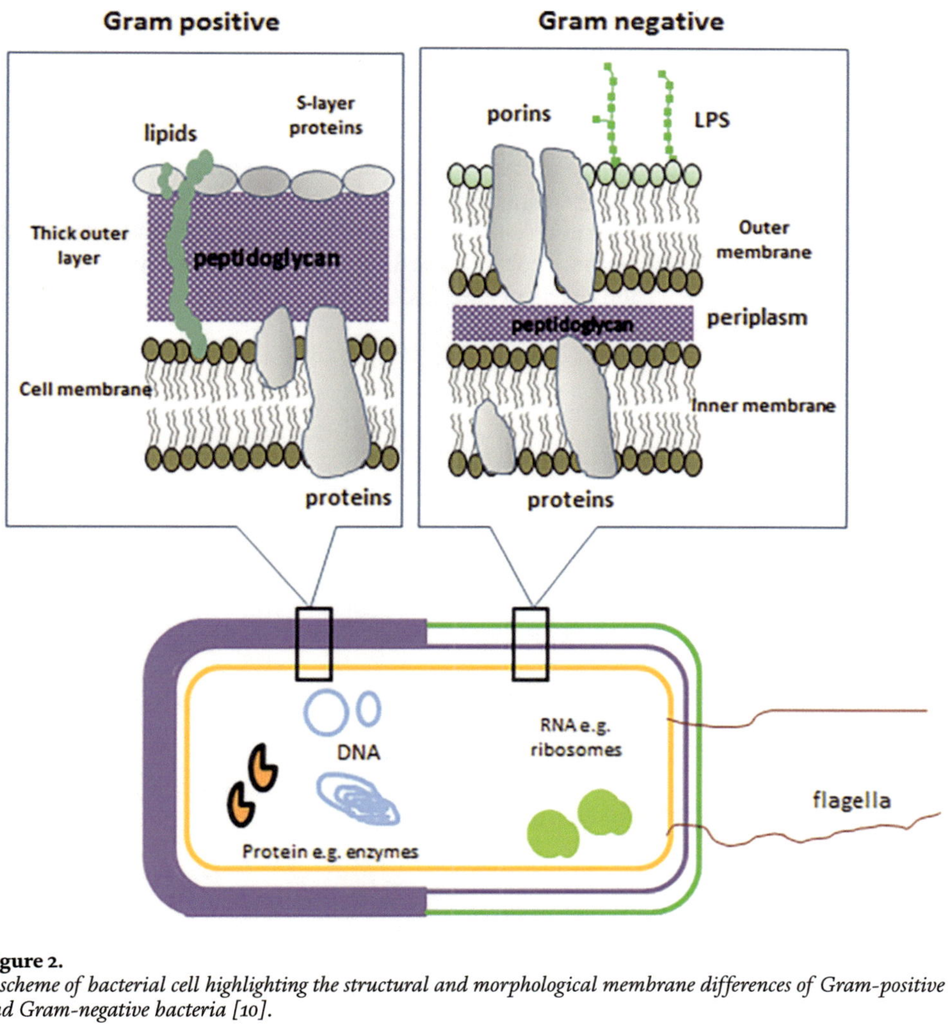

Figure 2.
A scheme of bacterial cell highlighting the structural and morphological membrane differences of Gram-positive and Gram-negative bacteria [10].

nucleic acids [10, 15–17]. These groups of amino acids, lipids and nucleic acids are the constituents of plasma, mitochondrial, lysosomal and nuclear membranes in bacterial and eukaryotic cells, as well as the viral envelope glycoproteins, coat proteins, host receptor recognition and binding proteins [10, 15–17].

Comparing the rate and extent of singlet oxygen inactivation of Gram-positive and Gram-positive bacteria, the Gram-positive bacteria are easily inactivated due to differences in the structure of their cell membrane [3, 4, 10, 18]. Gram-negative bacteria have an additional outer membrane apart from the cytoplasmic (inner) membrane (**Figure 2**) [10], giving them extra protection against antimicrobial agents including singlet oxygen and other ROS produced during photosensitisation.

For viruses, like other antimicrobial agents, the general photodynamic inactivation trend is that enveloped viruses are easily inactivated compared to non-enveloped viruses [10, 16, 17]. The virus attachment/host-recognition protein is thought to be most susceptible to the singlet oxygen-mediated oxidation, thereby preventing the virus from attaching to the host cell and ultimately counted as inactivated (**Figure 3**) [10, 16, 17]. In non-enveloped viruses, the capsid proteins including host-recognition proteins are the immediate targets of singlet oxygen-mediated oxidation (**Figure 3**) [10, 16, 17]. In enveloped viruses, the envelope glycoproteins including host-recognition proteins are the potential targets of singlet oxygen oxidation (**Figure 3**) [10, 16, 17]. It is believed that prolonged exposure of viral nucleic acid to singlet oxygen may also result in oxidative damage. Virus capsids may contain small pores even as they serve to protect the genome [10, 16, 17]. Also, the capsids of non-enveloped viruses exhibit a phenomenon called "capsid breathing" where under physiological conditions, the capsids undergo constant motion that suggests a dynamic state. Therefore, it has been suggested that through capsids small pores and breathing access to the viral genome could result in oxidation-induced damages to genome or the capsid even though the permeability of viral capsids to singlet oxygen molecules has not been shown [10, 16, 17].

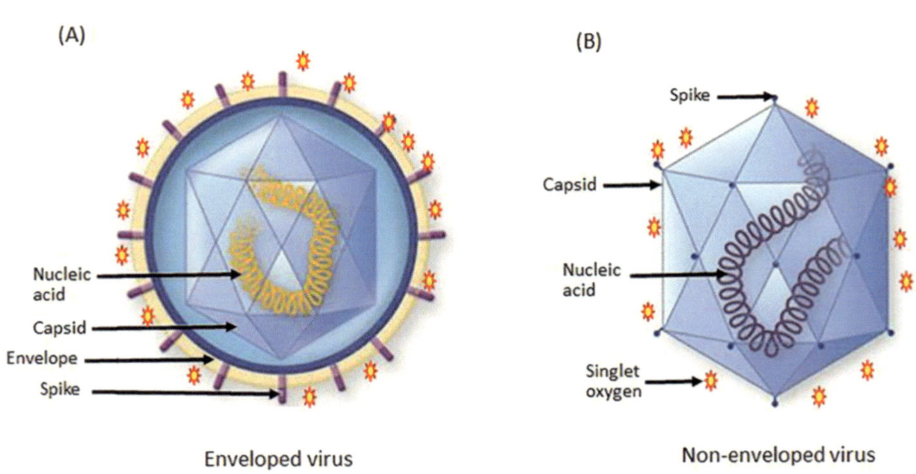

Figure 3.
Schemes of (A) Enveloped virus and (B) Non-enveloped virus showing first targets of singlet oxygen-mediated oxidation [10].

5. Food-grade dye extracts/photosensitisers from non-staple plants/plant parts

Food-grade photosensitisers are usually favoured and preferred, especially for food disinfection and sanitation [3–9]. The food-grade photosensitisers can be natural or synthetic and are Generally Regarded as Safe (GRAS) based on the rulings and approvals of food regulatory and standard organisations and or based on long history of use of the photosensitisers or their sources as food/medicine by humans/animals with no known reports of acute and chronic/cumulative toxicity [3–9]. The call for the alternative use of food-grade photosensitisers, especially for food disinfection and preservation, is to enhance food safety and security. Therefore, sourcing food-grade photosensitisers from non-staple food crops/plants is the ideal thing to do to avoid food security risks associated with the use of staple food crops/plants for things other than sustenance and consumption by humans and animals [3, 4]. Also, the food-grade photosensitisers should have a broad spectrum and high antimicrobial activity, be safe usage by humans as cosmetics/medicine/for making drinks and beverages and be available in abundance in many regions of the world [3, 4].

Natural food-grade dye extract/photosensitisers from non-staple plants can be refined or crude. The refined food-grade photosensitisers may mostly only exert photodynamic antimicrobial activity while the crude ones can exert both the photodynamic and non-photodynamic antimicrobial activity. The crude photosensitisers would be cheap, easy to get and use, and would have additional non-photodynamic antimicrobial activity, anti-inflammatory and antioxidation properties attributable to phytochemicals present in these plant extracts [3, 4]. Also, the dyes/photosensitisers and other phytochemicals in the plants with light or no light can have synergistic effects suitable for food disinfection and biomedical applications [3, 4].

5.1 Curcuminoids

Curcumin (**Figure 4**) is usually isolated from the rhizome/root of *Curcuma longa* (Turmeric) and has been shown to have a wide range of biological activities such as anti-inflammatory, antitumor, antiviral and antibacterial properties [8, 19–23]. Turmeric is a perennial plant native to Southeast Asia and Africa and has been used since ancient times as a spice, medicine, colouring and flavouring of food. It has broad

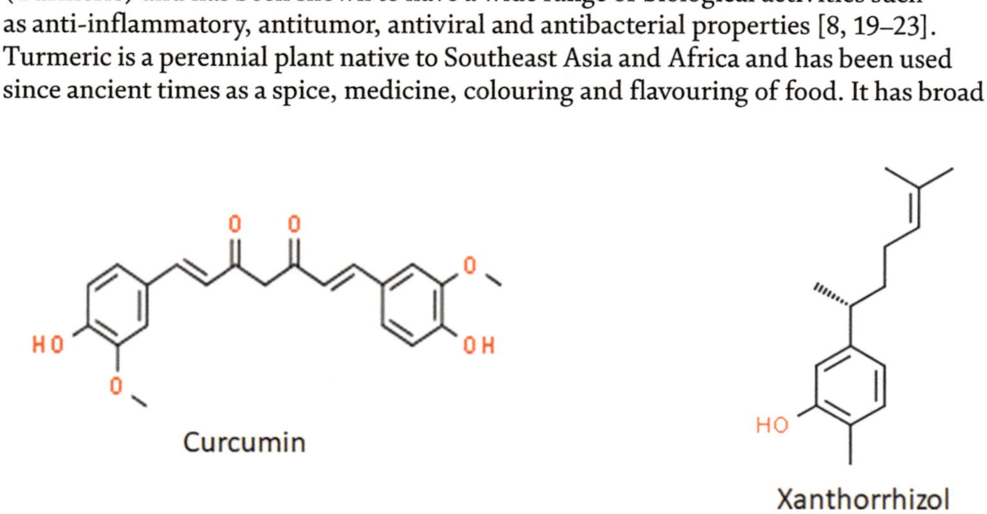

Figure 4.
Chemical structures of curcumin and xanthorrhizol downloaded from ChemSpider.

applications in photodynamic therapy and inactivation studies [8, 19–23]. Absorption wavelength of curcumin ranges from 405 to 435 nm [8]. Photodynamic inactivation ability of curcumin has been confirmed on Gram-positive, Gram-negative bacteria and *Candida* species in food disinfection including fresh produce and food-related diseases, periodontitis, wound healing, and multidrug resistance [8, 19–23]. It is, however, important to note that curcumin is not soluble in water and usually necessitates some modifications to improve on its solubility in water [8].

Crude extracts of *Curcuma longa* and *Curcuma xanthorrhiza* (Java ginger) were reported to have photodynamic therapy and inactivation properties, which were specifically used against periodontitis, adjunctive treatment, and planktonic biofilms upon exposure of 840 to 12,000 W/m^2 light intensities [8, 21]. Curcumin is the photosensitising agent in the extract of *Curcuma longa* while curcumin and xanthorrhizol (**Figure 4**) are the suspected photosensitisers in the extract of *Curcuma xanthorrhiza* [8, 21]. Although xanthorrhizol has almost the same physiological functions as curcumin, it is about 10 times more soluble in water than curcumin [8, 21]. The *Curcuma xanthorrhiza,* which has been used as a medicinal plant, food drink and spice, originates from Indonesia, specifically from Java Island and spread to other places such as Malaysia, Thailand, Philippines, China, Barbados, India, Japan, Korea, the United States and some countries in Europe [8, 21].

5.2 Alkaloids

Palmatine (**Figure 5**) is a protoberberine and photoactive quinolone-based alkaloid which can be isolated from several plants [8]. Some examples of plants with a long history of use in Chinese medicines and where palmatine can be found are *Phellodendron amurense, Coptis chinensis, Corydalis yanhusuo, Tinospora cordifolia, Tinospora sagittata, Phellodendron amurense* and *Stephania yunnanensis* [8, 24, 25]. The *Enantia chlorantha* extract has palmatine as its major constituent and is studied for use in the treatment of bacterial, viral and fungal infections, hypertension, jaundice, cancer, dysentery, inflammation and liver-related diseases. Palmatine hydrochloride has been shown to mediate the photodynamic therapy of breast cancer MCF-7 cells [8, 24, 25].

Berberine (**Figure 5**) is also a photoactive quinolone-based alkaloid that can be isolated from the roots, rhizomes, stems, and bark of several plants with a long history of use in traditional medicines, especially in China [8, 25–27]. Some of the plants where berberine can be found include *Berberis aristata, Berberis vulgaris, Berberis thunbergii, Hydrastis canadensis, Mahonia aquifolium, Xanthorhiza simplicissima, Coptis chinensis, Phellodendron amurense, Tinospora cordifolia, Eschscholzia californica* and *Argemone mexicana* [8, 25–27].

The yellow colour of berberine was used to dye wool, leather, and wood. The berberine possesses photodynamic therapy and inactivation properties which have been harnessed to treat several types of non-neoplastic diseases [8, 25–27]. The berberine coupled with a near ultraviolet (UV) light showed phototoxicity to mosquito larvae. The dye also has a phototoxicity effect on the HaCaT keratinocytes when irradiated with ultraviolet A (UVA). Photodynamic inactivation ability of berberine has been reported for both Gram-positive and Gram-negative bacteria [8, 25–27].

Pterins (**Figure 5**) can be found in most living organisms including bacteria, bluegreen algae, trypanosomes and mammals [8, 28–30]. The pterins have also been detected in the chloroplasts of spinach leaves and the cotyledons of pea sprouts

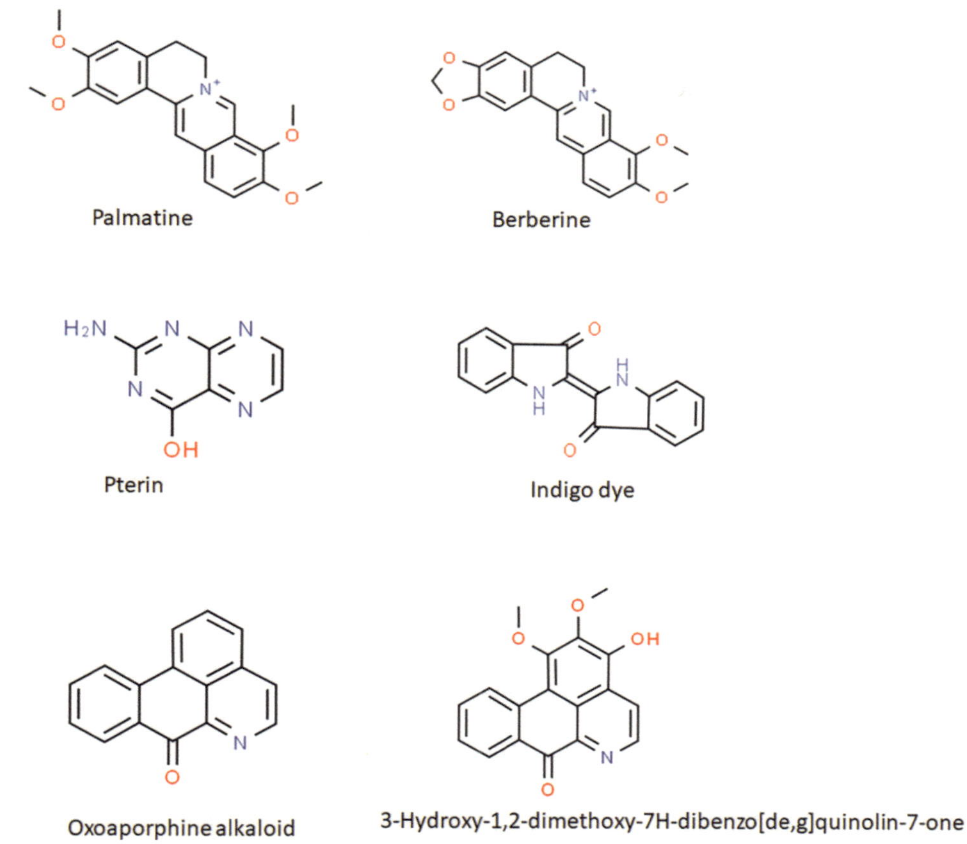

Figure 5.
Chemical structures of photoactive alkaloids downloaded from ChemSpider.

[8, 25, 31, 32]. Pterin photosensitisers have a good ability to generate singlet oxygen that can be harnessed for use in fine chemical synthesis, wastewater treatment and photodynamic applications such as sunlight-activated insecticides and herbicides, blood sterilisation, as well as photodynamic cancer therapy [8, 25, 31, 32]. The pterin-6-carboxylic acid with a 31 W/m^2 light intensity demonstrated effective photodynamic inactivation against *Klebsiella pneumonia* [32].

Indigo photosensitiser (indigo dye) (**Figure 5**) can be isolated from the plant *Indigofera tinctoria* [8, 28]. The *Indigofera tinctoria* has been naturalised in tropical and temperate regions of Asia and Africa, and it is known to be one of the sources of indigo dye, which shows photoreactivity against both the Gram-positive and Gram-negative bacteria [8, 28]. Crude methanol extract of *Indigofera truxillensis* was reported to have effective photodynamic antimicrobial activity against multidrug-resistant microorganisms including bacteria and Candida species [8, 28].

Crude extract of *Guatteria blepharophylla* has been used to control a broad range of microorganisms [8, 28]. It was reported that a 5-minute exposure of the extract to 900 W/m^2 light (660 nm) demonstrated reasonable photodynamic antimicrobial ability [8, 28]. The suspected photosensitising agents in the extract are two isoquinoline alkaloids – Oxoaporphine alkaloid and 3-Hydroxy-1,2-dimethoxy-7H-dibenzo[de,g] quinolin-7-one (**Figure 4**) [8, 28].

5.3 Anthraquinones

Photoactive anthraquinones (**Figure 6**) such as emodin, rhein, rubiadin, physcion, carminic acid and purpurin can be isolated from the root of the plant – *Polygonum cuspidatum* that had been extensively used in traditional Chinese medicine [8]. Aloe emodin-a photoactive anthraquinone is commonly used for the photodynamic inactivation of microorganisms [8, 29, 30]. Aloe-emodin has effective photodynamic inactivation ability against fungi and both Gram-positive and Gram-negative bacteria including *Staphylococcus aureus, Escherichia coli, Acetinobacter baumannii,* and *Candida albicans* [8, 29, 30].

Parietin (physcion) (**Figure 6**), which is also one of the anthraquinone photosensitiser is a predominant cortical pigment of lichens in the genus *Caloplaca* [33, 34]. The parietin is a secondary product of the lichen *Xanthoria parietina*, and it can also be found in the roots of *Rumex crispus* (curled dock) which is native to Europe and Western Asia [33, 34]. The leaves of curled dock have been used as food and sources of vitamins (A and C) and minerals (iron and potassium) [33, 34]. In Western traditional medicine, the root of curled dock is used to treat anaemia [33, 34]. It has been reported that parietin, when irradiated with a 1000 W/m^2 light intensity, demonstrated a strong photodynamic inactivation ability against both Gram-positive and Gram-negative bacteria [8, 33, 34].

5.4 Perylenequinones

Perylenequinones (**Figure 7**) can be sourced mainly from the plant – *Fagopyrum esculentum* (Herb buckwheat), an aphid – *Hormaphis* sp., and *Blepharisma* sp., which is a protozoan [8]. Hypocrellin A and its acid converted form hypocrellin B (**Figure 7**) can be isolated from *Hypocrella bambusae,* which is a parasitic fungus in bamboo [35–37]. The *Hypocrella bambusae* is commonly used in Chinese traditional medicine for the treatment of skin diseases, rheumatoid arthritis and gastric diseases [35–37]. Both hypocrellin A and hypocrellin B exhibited good photodynamic antimicrobial activity with the generation of large amounts of singlet oxygen [35–37]. Hypocrellins have been shown to have photodynamic antifungal activity against *Candida albicans* upon exposure to light [36].

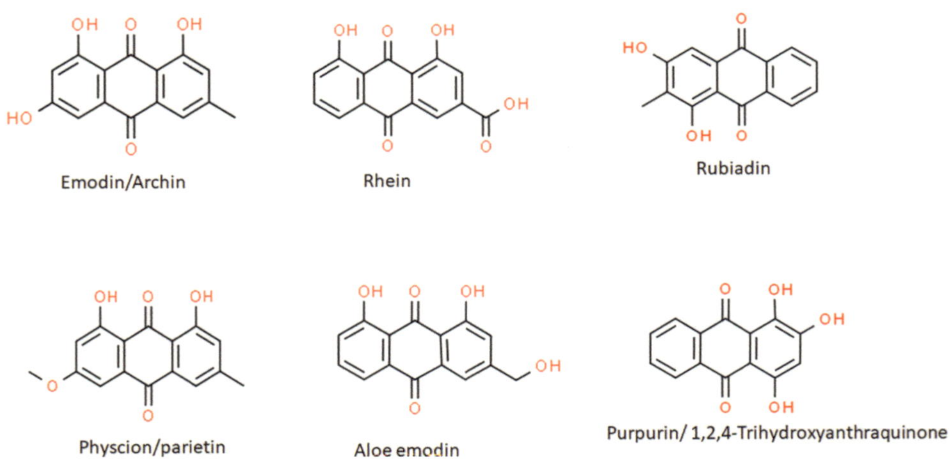

Figure 6.
Chemical structures of some photoactive anthraquinones of plant origins downloaded from ChemSpider.

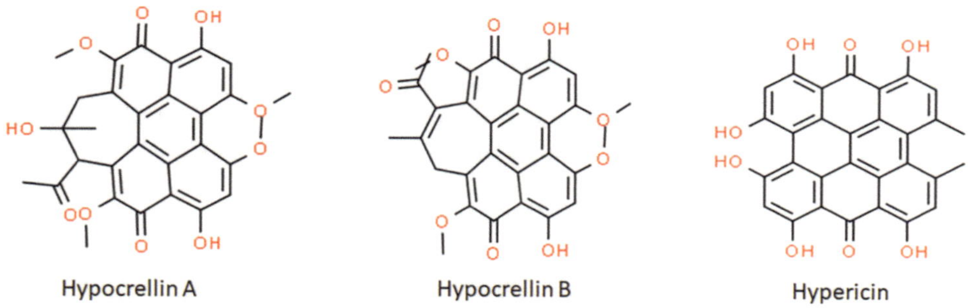

Figure 7.
Chemical structures of hypocrellins and hypericin downloaded from ChemSpider.

Hypericin-one of perylenequinone photosensitisers (**Figure 7**), is usually isolated from the plant *Hypericum perforatum* (Saint John's wort) [8, 38–41]. The Saint John's wort has been extensively used in folk medicine due to its antiviral, antitumor, antidepressant, and antibiotic properties [38–41]. The hypericin had demonstrated effective photodynamic inactivation of fungi due to its high quantum yield of reactive oxygen species and slow photobleaching [38–41]. The use of hypericin together with acetylcysteine exhibited a synergistic effect on *Staphylococcus aureus* biofilms [38]. Also, the photodynamic inactivation ability of hypericin has been exploited to combat food-borne pathogens and acne [8]. It is, however, important to note that hypericin has poor solubility in water and absorbs light at 590–600 nm [8]. Hypericin, when exposed to orange light (590 nm, 150 W/m^2) demonstrated a potent antimicrobial activity against the Gram-positive bacteria *Staphylococcus aureus* and Methicillin-Resistant *Staphylococcus aureus* (MRSA) and the fungus – *Candida albicans* [8]. Crude extracts of Saint John's wort exhibited photodynamic antimicrobial ability and had been shown to inhibit oral biofilms [39, 40]. The extract was also investigated against human enteric virus under solar irradiation with promising results [41].

5.5 Riboflavin

Riboflavin, also known as vitamin B2 (**Figure 8**), is a popular flavin that is present in many organisms including human tissues, plant leaves, eggs and mushrooms [8, 42, 43]. It has two absorption peaks in the UVA (360 nm) and blue (visible, 440 nm)

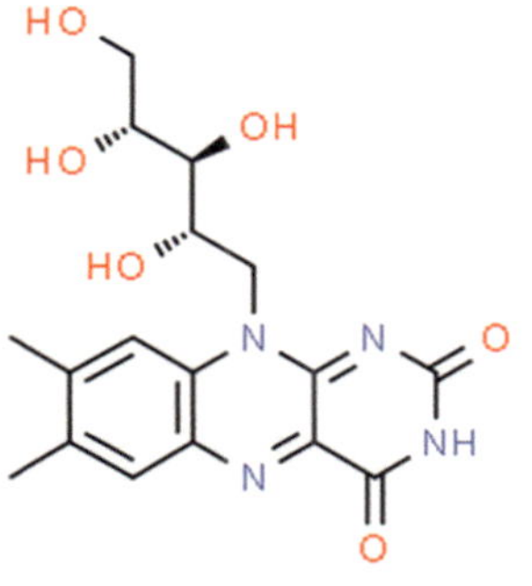

Figure 8.
Chemical structure of riboflavin downloaded from ChemSpider.

regions [8, 42, 43]. It has been used for photodynamic inactivation applications due to its high quantum yield of reactive oxygen species, thereby effectively inhibiting the growth of antibiotic-resistant bacteria such as enterohemorrhagic *Escherichia coli* and MRSA [8, 42, 43]. It has been shown that riboflavin, when exposed to light (460 nm, 150–580 W/m^2), successfully achieved inactivation of *Listeria monocytogenes* on smoked salmon fillets [8, 42, 43].

5.6 Chlorins

Chlorophyll, heme and cobalmin are natural molecules from plants and animals [8, 44–46]. They are tetrapyrrolic macrocycles and are considered as "pigments of life" [8, 44–46]. Chlorins (**Figure 9**) and Porphyrins photosensitisers are the most commonly studied photosensitisers for photodynamic therapy and inactivation applications [8, 44–46]. Chlorins are derived from chlorophyll a and have a maximum absorption ofetween 650 and 700 nm [8, 44–46]. Chlorins are also derivatives of porphyrins with just extra hydrogen atoms in the pyrrole ring [8, 44–46]. Chlorins, when exposed to light of appropriate wavelength, can generate large

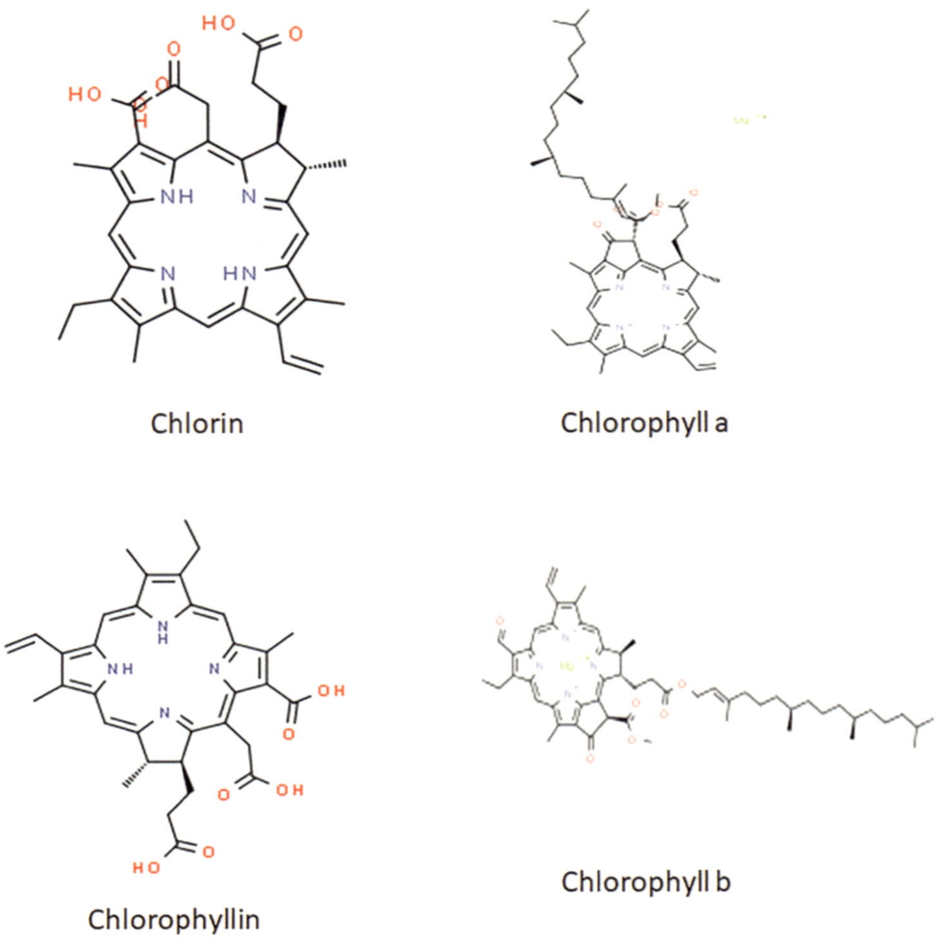

Figure 9.
Chemical structures of chlorin and chlorophylls downloaded from ChemSpider.

amount of singlet oxygen [8, 44–46]. They are insoluble in water and also unstable in the presence of light, acids and bases; these drawback limit their application [8, 44–46]. Many chlorins photosensitisers have however found applications in the microbial control of seeds, plant disease, and multidrug resistance [8, 44–46]. At low concentrations, chlorophyllin sodium salt (**Figure 9**) demonstrated effective photodynamic antimicrobial properties upon exposure to light intensities of 96 and 148 W/m^2 [8, 47, 48].

5.7 Carotenoids

Crude extracts of the plant – *Bixa orellana* have been reported to have effective photodynamic therapy and inactivation ability upon exposure to 1540 W/m^2 [8, 49]. The photosensitising agent in the extract is the dye – bixin and B-norbixin (**Figure 10**), which are apocarotenoids from the seeds of *Bixa orellana* that is native to South and Central America [8, 49]. The dye, also called annatto, is frequently used worldwide and have had applications not only in colouring foods, but also used in the textile, paint, and cosmetic industries as a colourant [8, 49]. It is the second most economically important dye after saffron. It has both medical and industrial importance [8, 49]. The leaves of *Bixa orellana* have been shown to have antimicrobial, anticonvulsant, analgesic and anti-inflammatory activities [8, 49]. Both the seeds and leaves of the plant have showed strong activity against Gram-positive and Gram-negative bacteria [8, 49]. Since the dye is natural and nontoxic, the extract of *Bixa orellana* extract could be considered as a potential photosensitiser for microbial control in foods and surface contacts [8, 49].

5.8 Monoterpenes and sesquiterpenes

Essential oils from *Boswelia carteri* (frankincense), *Origanum vulgare* (oregano), and thieves blend been shown to exhibit photodynamic antimicrobial activity [8, 50]. Monoterpenes and sesquiterpenes (**Figure 11**) are the major constituents of these essential oils, which have been reported to effectively control microorganisms in oral cavities through photodynamic inactivation [8, 50]. The *Boswelia carteri* – is a tree plant which is native to the countries in the south of the Arabian Peninsula including Oman and Yemen, and to Somalia, in the Horn of Africa [50]. Frankincense (Olibanum) is an aromatic resin that is usually harvested from *Boswelia carteri* and used in the production of incense and perfumes [50]. The *Origanum vulgare* is a flowering perennial woody plant native to the Mediterranean region and naturalised widely in the temperate Northern Hemisphere [50]. The flowers of *Origanum vulgare,* known as wild marjoram, are used as culinary herbs in Turkish, Greek, Spanish, Italian, Hispanic, and French cuisine [50]. The thieves blend is made up of essential oils from many medicinal plants and their parts including clove, lemon, cinnamon bark, *Eucalyptus radiata,* and Rosemary [8, 50].

Figure 10.
Chemical structures of bixin and B-norbixin downloaded from ChemSpider.

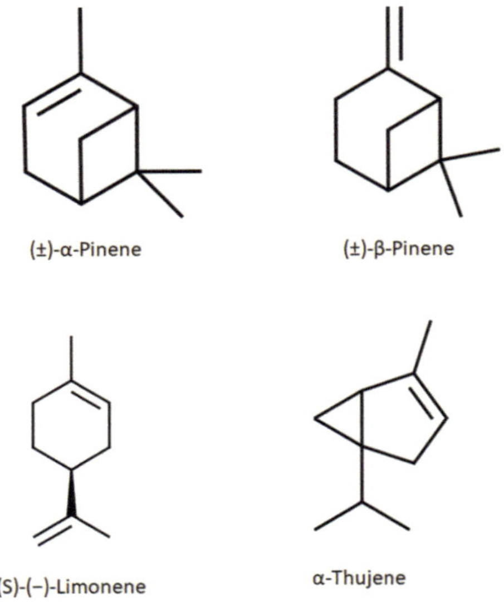

Figure 11.
Chemical structures of photoactive monoterpenes and sesquiterpenes downloaded from ChemSpider.

Figure 12.
General chemical structure of thiophene downloaded from ChemSpider.

5.9 Thiophenes

Crude extract from the plant – *Tagetes minuta* has been shown to have effective photodynamic inactivation activity against Candida species [8, 51]. Tagetes species including *Targetes minuta* are rich sources of natural thiophenes (**Figure 12**), which have several bioactivities including their antifungal activity which is enhanced by irradiation with UVA, thereby suggesting that their extracts could be used in photodynamic therapy and inactivation [8, 51]. Several essential oils are usually sourced from Tagetes species and used for food flavouring while the orange-yellow colours of their flowers are used for food colouring [51]. *Tagetes minuta* is an annual perennial herb that grows from temperate to tropical regions of the world [51]. The plant is known to have antimicrobial properties including antiseptic, antihelminthic and antifungal properties, thereby providing the basis for its usage as medicine for respiratory, digestive, and skin ailments and for the treatment of urinary tract infections [51].

Crude hexane extract of *Porophyllum obscurum* has been shown to have antifungal activity under UVA light [8, 52]. The extract is effective against oropharyngeal candidiasis [8, 52]. The four (4) thiophenes in the extract are thought to be responsible for their photoantifungal activity. The are 2, 2′:5′2″terthiophene, 5-(3-buten-1-ynyl)-2, 2′-bithiophene, 5-(4-acetoxy-1-butenyl)-2,2′- bithiophene and 5-(4-hydroxy-1-butenyl)-2,2′- bithiophene [8, 52]. The *Porophyllum obscurum* and other species spread from the United States, Mexico, Central America, Antilles, and South America [8, 52]. It is a perennial plant and is used in folk medicine

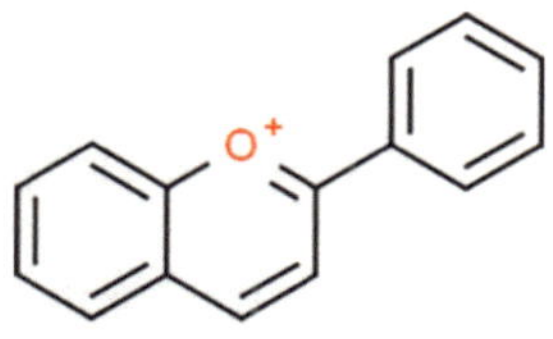

Figure 13.
Chemical structure of anthocyanin downloaded from ChemSpider.

for its activity against cramps and venereal diseases and for its diaphoretic and antispasmodic properties [52]. Several studies have reported that the plant and related species have antibacterial, anti-inflammatory and insecticide properties [52].

5.10 Anthocyanins

Crude methanol extract of calyces of *Hibiscus sabdariffa* (Hibiscus) has been reported to have photodynamic antibacterial against *Escherichia coli* (Gram-negative bacteria) and *Bacillus subtilis* (Gram-positive bacteria) [3]. *Hibiscus sabdariffa* is an annual tropical and subtropical bushy shrub with red or pale green stems and red or pale yellow inflated edible calyces [3]. It is native to West Africa but is now grown in other tropical and subtropical countries, especially in Africa and Asia [3]. The red pigmentation of the calyces is due to the high concentration of anthocyanins (**Figure 13**), which are thought to be the photosensitising agents in the extract [3]. Other chemical constituents of the calyces are organic acids, flavonoids, and other polyphenols [3]. The calyces of Hibiscus are used as food and for the production of beverage drinks [3].

6. Photostability and photodynamic inactivation characteristics of dye extracts and photosensitisers in different solutions/buffers

One of the desired good qualities of a photosensitiser/dye extract is photostable and not bleach upon prolonged exposure to light [10]. It is thought that bleaching of a photosensitiser upon exposure to light will lead to a decrease in the generation of reactive oxygen species either by type 1 photoreactions that can lead to the production of superoxide, peroxide and hydroxyl radical and ion or by type 2 photo reaction that can lead to production of singlet oxygen [3, 10]. The decrease in the generation of reactive oxygen species will lead to decrease or cessation of photodynamic therapy and the inactivation ability of the photosensitiser [3, 10]. The two main factors that determine whether a photosensitiser is photostable or bleaches upon exposure to light are chemical nature of the photosensitiser/dye extract and the medium solution/solvent type/composition [3, 10]. It is very possible for a photosensitiser to be photostable in one solution and bleach in another solution; this therefore necessitates the testing of dye extracts/photosensitisers in different solutions and buffers in order to harness their photodynamic and non-photodynamic antimicrobial ability for food disinfection and sanitation as well for some biomedical applications.

A study reported the use of four different solutions to test the photostability and photodynamic antibacterial activity of methanol extract of calyces of *Hibiscus sabdariffa* whose photosensitising agent is thought to be anthocyanins [3]. The four solutions used were distilled water (dH_2O), phosphate buffer saline (PBS) (pH 7.2), acetate-HCl (AH) Buffer (pH 4.6) and Tris base–HCl (TBH) buffer (pH 8.6). The photostability and bleaching study showed that the dye extract-*Hibiscus sabdariffa* (HS) is relatively photostable in dH_2O but bleaches in buffers-AH, TBH and PBS [3]. Over the 20 minutes of exposure to light, the HS in dH_2O only bleaches by 10.70% [3]. The rate of bleaching is higher in AH compared to in TBH and PBS [3]. The HS in AH beaches by 76.17% after 20 minutes of exposure to light while in the same dye in TBH and PBS, the bleaching observed were 51.76 and 42.86%, respectively [3]. The bleaching correlated with the extent of photodynamic inactivation observed in these solutions [3]. The bleaching can be linked to pH and/or buffer type/compositions [3]. And for these reasons, the application of the dye extract as photosensitiser maybe limited to only environmental application such as sanitation of fresh produce sanitation, where fresh water would be used as wash waters containing the dye extracts to photodynamically inactivate pathogenic and spoilage microorganisms [3]. However, for biological samples and solutions of high salts, the dye extract as photosensitiser may not be suitable for the photodynamic inactivation of microorganisms [3].

The dye extract in dH_2O demonstrated effective photodynamic inactivation of *Escherichia coli* (Gram-negative) and *Bacillus subtilis* (Gram-positive). This follows and may be linked to the photostability of the dye extracts in dH_2O. A photostable photosensitiser molecule at the ground state absorbs light and ultimately gets excited to a triplet state where it interacts with the molecular oxygen to generate ROS. The singlet oxygen and other ROS are cytotoxic and can inactivate microorganisms including bacteria in solution [3].

After the interaction of the triplet excited state of photostable photosensitisers with the oxygen, the molecules of the photosensitisers will go to the ground state to be excited again with the illuminated light, and the process continues with the resultant production of ROS. However, for non-photostable photosensitisers including the dye extract of calyces of Hibiscus in phosphate buffer saline (pH 7.2), acetate-HCl buffer (pH 4.6) and tris base–HCl buffer (pH 8.6), at each interactions of the excited triplet state of the molecules with molecular oxygen, the singlet oxygen and other ROS produced consumed the molecules of the photosensitisers that produces them thereby leading to the photodegredation and bleaching of the photosensitisers/dyes [3]. This led to a decrease in the molecules of the photosensitisers which corresponds to a decrease in the ROS generation [3]. As this continued, the ROS generation in solution would eventually stop within seconds or minutes. There will be insufficient ROS produced in the solution, and those produced degrade the molecules of photosensitisers that produce them instead of oxidising the bacteria in the solution; these maybe possible reasons for the lack or less bacteria photodynamic inactivation observed in phosphate buffer saline (pH 7.2), acetate-HCl buffer (pH 4.6) and tris base–HCl buffer (pH 8.6) [3].

7. Conclusions and future perspectives

There are many types of food-grade photosensitisers and dye extracts that can be sourced abundantly from non-staple plant extracts that have a long history of safe use as medicines and foods by humans. These photosensitisers of plant origin are diverse groups of compounds including curcumoids, alkaloids, anthraquinones, perylenequinones, flavins, chlorins, thiophenes, carotenoids, terpenes and anthocyanins. In this era of increasing microbial drug resistance and the quest to go green, especially in food disinfection and sanitation, the photodynamic and non-photodynamic antimicrobial ability of these potential food-grade photosensitisers could be harnessed and used to control spoilage and pathogenic microorganisms in foods and contact surfaces. This can serve as an alternative to complex and expensive conventional means of controlling microorganisms and has the potential of replacing it. It is very possible to use abundant sunlight to irradiate the photosensitisers to generate ROS since most of the photosensitisers can absorb in the visible region of the electromagnetic spectrum, thereby making food disinfection and sanitation possible even in places with no man-made energy/power sources/supply.

Acknowledgements

This publication was possible thanks to the SG-NAPI award supported by the German Ministry of Education and Research, BMBF through UNESCO – TWAS to Hussaini Majiya-SG-NAPI 4500474983.

Conflict of interest

The authors declare no conflict of interest.

Author details

Hussaini Majiya[1,2,3,4]

1 Department of Microbiology, Ibrahim Badamasi Babangida University, Lapai, Nigeria

2 Center for Applied Sciences and Technology Research, Ibrahim Badamasi Babangida University, Lapai, Nigeria

3 Trans-Saharan Disease Research Center, Ibrahim Badamasi Babangida University, Lapai, Nigeria

4 Directorate of Research and Development, Ibrahim Badamasi Babangida University, Lapai, Nigeria

*Address all correspondence to: husssainimajiya@ibbul.edu.ng

References

[1] de Oliveira EF, Tosati JV, Tikekar RV, Monteiro AR, Nitin N. Antimicrobial activity of curcumin in combination with light against *Escherichia coli* O157: H7 and *Listeria innocua*: Applications for fresh produce sanitation. Postharvest Biology and Technology. 2018;**137**:86-94

[2] Ghate VS, Zhou W, Yuk HG. Perspectives and trends in the application of photodynamic inactivation for microbiological food safety. Comprehensive Reviews in Food Science and Food Safety. 2019;**18**(2):402-424

[3] Majiya H, Galstyan A. Dye extract of calyces of *Hibiscus sabdariffa* has photodynamic antibacterial activity: A prospect for sunlight-driven fresh produce sanitation. Food Science & Nutrition. 2020;**8**(7):3200-3211

[4] Majiya H, Galstyan A. Potential of a methanolic extract of *Lawsonia inermis* (L.) leaf as an alternative sanitiser in the time of COVID-19 pandemic and beyond. Journal of Herbal Medicine. 2023;**38**:100633

[5] Zhu S, Song Y, Pei J, Xue F, Cui X, Xiong X, et al. The application of photodynamic inactivation to microorganisms in food. Food Chemistry: X. 2021;**12**:100150

[6] Ponzio RA, Ibarra LE, Achilli EE, Odella E, Chesta CA, Martínez SR, et al. Sweet light o'mine: Photothermal and photodynamic inactivation of tenacious pathogens using conjugated polymers. Journal of Photochemistry and Photobiology B: Biology. 2022;**234**: 112510

[7] Foresto E, Gilardi P, Ibarra LE, Cogno IS. Light-activated green drugs: How we can use them in photodynamic therapy and mass-produce them with biotechnological tools. Phytomedicine Plus. 2021;**1**(3):100044

[8] Polat E, Kang K. Natural photosensitizers in antimicrobial photodynamic therapy. Biomedicine. 2021;**9**(6):584

[9] Winter S, Tortik N, Kubin A, Krammer B, Plaetzer K. Back to the roots: Photodynamic inactivation of bacteria based on water-soluble curcumin bound to polyvinylpyrrolidone as a photosensitizer. Photochemical & Photobiological Sciences. 2013;**12**(10): 1795-1802

[10] Majiya HM. Photosensitiser functionalised nanofiber fabric for efficient light driven water disinfection [thesis]. England: University of Leeds; 2017

[11] DeRosa MC, Crutchley RJ. Photosensitized singlet oxygen and its applications. Coordination Chemistry Reviews. 2002;**233**:351-371

[12] Alves E, Costa L, Carvalho CM, Tomé JP, Faustino MA, Neves MG, et al. Charge effect on the photoinactivation of gram-negative and gram-positive bacteria by cationic meso-substituted porphyrins. BMC Microbiology. 2009;**9**:1-13

[13] Costa DC, Gomes MC, Faustino MA, Neves MG, Cunha A, Cavaleiro JA, et al. Comparative photodynamic inactivation of antibiotic resistant bacteria by first and second generation cationic photosensitizers. Photochemical & Photobiological Sciences. 2012;**11**: 1905-1913

[14] Costa L, Faustino MA, Tome JP, Neves MG, Tome AC, Cavaleiro JA, et al. Involvement of type I and type II

mechanisms on the photoinactivation of non-enveloped DNA and RNA bacteriophages. Journal of Photochemistry and Photobiology B: Biology. 2013;**120**:10-16

[15] Davies MJ. Singlet oxygen-mediated damage to proteins and its consequences. Biochemical and Biophysical Research Communications. 2003;**305**(3):761-770

[16] Majiya H, Adeyemi OO, Herod M, Stonehouse NJ, Millner P. Photodynamic inactivation of non-enveloped RNA viruses. Journal of Photochemistry and Photobiology B: Biology. 2018;**189**:87-94

[17] Majiya H, Adeyemi OO, Stonehouse NJ, Millner P. Photodynamic inactivation of bacteriophage MS2: The A-protein is the target of virus inactivation. Journal of Photochemistry and Photobiology B: Biology. 2018;**178**:404-411

[18] Majiya H, Chowdhury KF, Stonehouse NJ, Millner P. TMPyP functionalised chitosan membrane for efficient sunlight driven water disinfection. Journal of Water Process Engineering. 2019;**30**:100475

[19] Saitawee D, Teerakapong A, Morales NP, Jitprasertwong P, Hormdee D. Photodynamic therapy of Curcuma longa extract stimulated with blue light against Aggregatibacter actinomycetemcomitans. Photodiagnosis and Photodynamic Therapy. 2018;**22**:101-105

[20] Hormdee D, Rinsathorn W, Puasiri S, Jitprasertwong P. Anti-early stage of bacterial recolonization effect of curcuma longa extract as photodynamic adjunctive treatment. International Journal of Dentistry. 2020;**2020**:1-7

[21] Lee HJ, Kang SM, Jeong SH, Chung KH, Kim BI. Antibacterial photodynamic therapy with curcumin and *Curcuma xanthorrhiza* extract against *Streptococcus mutans*. Photodiagnosis and Photodynamic Therapy. 2017;**20**:116-119

[22] Salleh NAM, Ismail S, Ab Halim MR. Effects of *Curcuma xanthorrhiza* extracts and their constituents on phase II drug-metabolizing enzymes activity. Pharmacognosy Research. 2016;**8**(4):309

[23] Kubrak TP, Kołodziej P, Sawicki J, Mazur A, Koziorowska K, Aebisher D. Some natural photosensitizers and their medicinal properties for use in photodynamic therapy. Molecules. 2022;**27**(4):1192

[24] Wu J, Xiao Q, Zhang N, Xue C, Leung AW, Zhang H, et al. Palmatine hydrochloride mediated photodynamic inactivation of breast cancer MCF-7 cells: Effectiveness and mechanism of action. Photodiagnosis and Photodynamic Therapy. 2016;**15**:133-138

[25] Siewert B, Stuppner H. The photoactivity of natural products–an overlooked potential of phytomedicines? Phytomedicine. 2019;**60**:152985

[26] An YW, Jin HT, Yuan B, Wang JC, Wang C, Liu HQ. Research progress of berberine mediated photodynamic therapy. Oncology Letters. 2021;**21**(5):1-10

[27] Andreazza NL, Vevert-Bizet C, Bourg-Heckly G, Sureau F, Salvador MJ, Bonneau S. Berberine as a photosensitizing agent for antitumoral photodynamic therapy: Insights into its association to low density lipoproteins. International Journal of Pharmaceutics. 2016;**510**(1):240-249

[28] Andreazza NL, de Lourenço CC, Stefanello MÉA, Atvars TDZ, Salvador MJ. Photodynamic

antimicrobial effects of bis-indole alkaloid indigo from *Indigofera truxillensis* Kunth (Leguminosae). Lasers in Medical Science. 2015;**30**:1315-1324

[29] Wang Y, Li J, Geng S, Wang X, Cui Z, Ma W, et al. Aloe-emodin-mediated antimicrobial photodynamic therapy against multidrug-resistant *Acinetobacter baumannii*: An *in vivo* study. Photodiagnosis and Photodynamic Therapy. 2021;**34**:102311

[30] Ma W, Liu C, Li J, Hao M, Ji Y, Zeng X. The effects of aloe emodin-mediated antimicrobial photodynamic therapy on drug-sensitive and resistant *Candida albicans*. Photochemical & Photobiological Sciences. 2020;**19**: 485-494

[31] Buglak AA, Telegina TA, Vorotelyak EA, Kononov AI. Theoretical study of photoreactions between oxidized pterins and molecular oxygen. Journal of Photochemistry and Photobiology A: Chemistry. 2019;**372**:254-259

[32] Tosato MG, Schilardi P, de Mele MFL, Thomas AH, Lorente C, Miñán A. Synergistic effect of carboxypterin and methylene blue applied to antimicrobial photodynamic therapy against mature biofilm of *Klebsiella pneumoniae*. Heliyon. 2020;**6**(3):e03522

[33] Varol M. Parietin as an efficient and promising anti-angiogenic and apoptotic small-molecule from *Xanthoria parietina*. Revista Brasileira de Farmacognosia. 2020;**29**:728-734

[34] Comini LR, Morán Vieyra FE, Mignone RA, Páez PL, Laura Mugas M, Konigheim BS, et al. Parietin: An efficient photo-screening pigment *in vivo* with good photosensitizing and photodynamic antibacterial effects *in vitro*. Photochemical & Photobiological Sciences. 2017;**16**:201-210

[35] Xiao Q, Wu J, Pang X, Jiang Y, Wang P, Leung AW, et al. Discovery and development of natural products and their derivatives as photosensitizers for photodynamic therapy. Current Medicinal Chemistry. 2018;**25**(7):839-860

[36] Yang Y, Wang C, Zhuge Y, Zhang J, Xu K, Zhang Q, et al. Photodynamic antifungal activity of hypocrellin A against *Candida albicans*. Frontiers in Microbiology. 2019;**10**:465191

[37] Jia Q, Zheng X, Ge J, Liu W, Ren H, Chen S, et al. Synthesis of carbon dots from *Hypocrella bambusae* for bimodel fluorescence/photoacoustic imaging-guided synergistic photodynamic/photothermal therapy of cancer. Journal of Colloid and Interface Science. 2018;**526**:302-311

[38] Kashef N, Karami S, Djavid GE. Phototoxic effect of hypericin alone and in combination with acetylcysteine on *Staphylococcus aureus* biofilms. Photodiagnosis and Photodynamic Therapy. 2015;**12**(2):186-192

[39] Jendželovská Z, Jendželovský R, Kuchárová B, Fedoročko P. Hypericin in the light and in the dark: Two sides of the same coin. Frontiers in Plant Science. 2016;**7**:191078

[40] Vollmer A, Al-Ahmad A, Argyropoulou A, Thurnheer T, Hellwig E, Attin T, et al. Antimicrobial photoinactivation using visible light plus water-filtered infrared-A (VIS+ wIRA) and hypericum perforatum modifies *in situ* oral biofilms. Scientific Reports. 2019;**9**(1):20325

[41] Ryberg EC, Knight J, Kim JH. Farm-to-tap water treatment:

Naturally-sourced photosensitizers for enhanced solar disinfection of drinking water. ACS ES & T Engineering. 2020;**1**(1):86-99

[42] Maisch T. Photoantimicrobials—An update. Translational Biophotonics. 2020;**2**(1-2):e201900033

[43] Josewin SW, Ghate V, Kim MJ, Yuk HG. Antibacterial effect of 460 nm light-emitting diode in combination with riboflavin against *Listeria monocytogenes* on smoked salmon. Food Control. 2018;**84**:354-361

[44] Dias LD, Mfouo-Tynga IS. Learning from nature: Bioinspired chlorin-based photosensitizers immobilized on carbon materials for combined photodynamic and photothermal therapy. Biomimetics. 2020;**5**(4):53

[45] Pucelik B, Sułek A, Dąbrowski JM. Bacteriochlorins and their metal complexes as NIR-absorbing photosensitizers: Properties, mechanisms, and applications. Coordination Chemistry Reviews. 2020;**416**:213340

[46] Walter AB, Simpson J, Jenkins JL, Skaar EP, Jansen ED. Optimization of optical parameters for improved photodynamic therapy of *Staphylococcus aureus* using endogenous coproporphyrin III. Photodiagnosis and Photodynamic Therapy. 2020;**29**:101624

[47] Žudyté B, Lukšiené Ž. Toward better microbial safety of wheat sprouts: Chlorophyllin-based photosensitization of seeds. Photochemical & Photobiological Sciences. 2019;**18**:2521-2530

[48] Glueck M, Hamminger C, Fefer M, Liu J, Plaetzer K. Save the crop: Photodynamic inactivation of plant pathogens I: Bacteria. Photochemical & Photobiological Sciences. 2019;**18**: 1700-1708

[49] Gonçalves MLL, da Mota ACC, Deana AM, de Souza Cavalcante LA, Horliana ACRT, Pavani C, et al. Antimicrobial photodynamic therapy with *Bixa orellana* extract and blue LED in the reduction of halitosis—A randomized, controlled clinical trial. Photodiagnosis and Photodynamic Therapy. 2020;**30**:101751

[50] Dascalu LM, Moldovan M, Prodan D, Ciotlaus I, Popescu V, Baldea I, et al. Assessment and characterization of some new photosensitizers for antimicrobial photodynamic therapy (aPDT). Materials. 2020;**13**(13):3012

[51] Giacone L, Cordisco E, Garrido MC, Petenatti E, Sortino M. Photodynamic activity of *Tagetes minuta* extracts against superficial fungal infections. Medical Mycology. 2020;**58**(6):797-809

[52] Postigo A, Funes M, Petenatti E, Bottai H, Pacciaroni A, Sortino M. Antifungal photosensitive activity of *Porophyllum obscurum* (Spreng.) DC.: Correlation of the chemical composition of the hexane extract with the bioactivity. Photodiagnosis and Photodynamic Therapy. 2017;**20**:263-272

Chapter 4

Potent Natural Dyestuffs from Onion Skin to Fabrics Using or Lacking Bio-Mordants

Sidra Mubeen, Tahira Batool and Ghulam Sabir Hussain

Abstract

Natural dyes yield a remarkable range of items with complex colouration that go well together. Skin that constitutes the outermost part of onion contains certain phytochemicals, such as; cardiac glycosides, tannins, flavonoids (quercetin, etc.), anthocyanidins (pelargonidin, etc.), natural dyestuffs, and reducing sugars in aqueous medium. Among the various phytochemicals, tannins, anthraquinones, and flavonoids are crucial for giving cotton fabric its shade and vibrancy. Wool has been dyed in the past and present using onion peel. It has historically been utilised in rugs and carpets in Anatolia, Turkey. It can be applied both with and without bio-mordants to silk fabric. Tea leaves, tamarind, aloe vera, and ash from acacia bark are utilised as bio-mordants. Furthermore, colour measurement values, such as Tristimulus characteristics, Lab values, and munsell notations, are applied to dyed materials.

Keywords: natural dyestuff, onion dye, natural colourant, bio-mordants, fabrics dyeing

1. Introduction

Colourants called dyes and pigments are used in food, medicine, textiles, and other essential items that are necessary for human survival [1]. Because dyes are more soluble and contain finer particles than pigments, they vary from pigments in this regard [2]. The use of colourants dates back to the Stone Age, and many ancient civilizations expanded their use of dyes to textiles once weaving techniques advanced [3]. Evidence demonstrates that the use of pigments and dyes dates back to the dawn of human civilization; natural pigments were employed by the ancient Egyptian, Roman, Greek, and Indian civilizations [4]. Among the popular ancient colours were madder, blue indigo, and yellow made from turmeric or saffron; crucial pigments included ochre, limestone, and charcoal [5]. Animals, minerals, and plants were the main sources of natural colourants. Almost every component of the plant, including the leaves, roots, fruit, flowers, bark, wood, and seeds, was used to create a variety of colours and their combinations [6, 7]. Until the middle of the nineteenth century, when William Henry Perkin created the first synthetic dye, mauve or aniline purple, in 1856, there was no interruption to the use of natural pigments and dyes [8, 9]. Dyes have a variety of uses; for example, fluorescent natural dyes can be successfully employed in beetroot cell imaging [10]. The use of natural dyes is a promising advancement in the field of dye-sensitised solar cells (DSSCs). Biosensors (indigo), dye-sensitised solar cells, extract used for cancer treatment (*Punica granatum*), and corrosion inhibitors (curcumin, lawson, and berberine derived from plant materials) contain nitrogen (N), sulphur (S), and O heteroatoms in their structure [1].

Using a mordant is usually required throughout the natural dyeing process. Metallic salts including sulphates, chlorides, hydroxides, and oxides can be used as a source of mordant. Alum and iron are two of the metal mordants used in natural dyes that are harmless to the environment and are not subject to any ecoregulations [11]. Numerous of these mordants have the potential to contaminate soil and produce waste in the form of water [12]. The incompatibility of these mordants with natural dyeing's eco-friendly tenets presents another issue. Widespread use would result in significant environmental issues [13, 14].

However, bio-mordants, which have enough colouring and firmness properties, are said to be environmentally friendly and sustainable substitutes for metal mordants [15]. Sources of bio-mordant materials include hyperaccumulative metal plants and plants high in tannin. A number of bio-mordants, including pomegranate peel, rosemary, and thuja leaves, have been suggested as viable substitutes for potassium dichromate, stannous chloride, iron sulphate II, copper sulphate II, and aluminium [16]. Plant parts that contain natural polyphenols, commonly referred to as tannins, include bark, wood, fruits, fruit peels, leaves, roots, and plant galls [17].

2. Onion skin as a natural dyestuff

One cuisine item that is typically seen on tables in Asian nations is onions. In addition to having high-quality ingredients, it can be claimed that this product remains a fixture in both dinners and salads [18]. The onion skins are wasted after use. At the grocery shop, onions (*Allium cepa*) come in three primary varieties based on their colour; white, yellow, and red [19], as shown in (**Figure 1**). However, appropriate names such as Red Zeppelins or Australian Browns are seen at farmers' markets. Red is usually the sharpest, yellow is in the middle, and white is typically the mildest in terms of pungency [20].

The edible part of an onion (*Allium cepa*) is the swelling bases of the leaves, which comprise the flesh [21, 22]. A well-ripened onion has dried out and lost much of its greenery, leaving just a little amount of shrivelled tissue at the ends of the outer fleshy scales [23]. The world's dry onion production region is in an exceptional position.

2.1 Chemical composition of onion skin

A variety of onion waste components, including ash, protein, carbohydrates, and total dietary fibre, are excellent sources of several minerals, including potassium, calcium, magnesium, iron, zinc, manganese, selenium, total phenolics, flavonoids, and flavonols [22, 24]. Additionally, the powdered onion skin revealed that it had a considerable amount of ash, total dietary fibre, soluble dietary fibre, and insoluble dietary fibre but less protein and crude fat [25]. Onion outer scales were found to include ash, moisture, carbohydrates, crude protein, crude fibre, and crude fat [26]. Additionally, the study identified

Figure 1.
Different types of onions.

12 distinct fatty acids from the onion peel's outer layer, of which half were unsaturated and the other half saturated [27]. The fatty acid profile showed that the total amount of saturated fatty acids was 21.42%, while the total amount of unsaturated fatty acids was 76.79%. Lauric acid (0.94%), myristic acid (1.28%), palmitic acid (9.80%), palmitoleic acid (2.84%), stearic acid (8.81%), oleic acid (17.57%), linoleic acid (52.87%), linolenic acid (2.88%), arachidic acid (0.59%), behenic acid (1.23%), erucic acid (0.63%), and lignoceric acid (0.54%) were among the fatty acid profile [28, 29]. Research has also shown that the peel of onions, which mostly consists of the phenolic component quercetin, is a more concentrated source of phytochemicals than the edible flesh. Furthermore, onion skin has been found to contain quercetin 3,4′-diglycoside, protocatechuic acid, kaempferol, 2-(3,4-dihydroxybenzoyl)-2,4,6-trihydroxy- 3(*2H*)-benzofuranone, isorhamnetin, quercetin-7,4′-diglycoside, isorhamnetin-3,4'diglycoside, quercetin-3-glycoside, quercetin-4′-glucoside, isorhamnetin-4′-glycoside, protocatechoyl quercetin, quercetin dimer 4′-glycoside, quercetin dimer hexoside, and quercetin trimer [30]. India produces over 22.43 million tonnes of onions annually, making it the second-largest producer of onions worldwide. Since onion peels are basically removed before being used, the yield of usable peeled onions varies from 73.5 to 81.6% depending on the onion's size, which ranges from 2.5 to 4.5 inches. As a result, on average, onion peels make up between 10 and 25% (%w/w basis) of the onion's total weight, producing a sizable amount of vegetable waste material. Based on a further extrapolation of the data, India produces between 2.3 and 4.5 million tonnes of onion peels annually. This would lower the economic worth of the precious onion material as an agricultural commodity, in addition to creating a significant environmental issue with trash management. Because of this, developing nations like India have a lot of opportunity to develop some value-added and environmentally benign products, which are essential for creating jobs and economic growth, in addition to developing environmentally benign products.

This research examines the practical application of onion peels for advantageous purposes, including the extraction of natural dyes. The plant's outer skins have long been employed in Turkey's textile industry as a natural colouring agent. Because of this, natural dyes made from these vegetable wastes can provide economic benefit and create jobs. Therefore, hand-woven carpets and rugs could benefit from using these wastes for dyeing [31].

2.2 Extraction methods for obtaining dye

The extraction of dyes from onion skin has been studied in various ways; measured amounts of ground-up samples (10 g) were wrapped in calico and put inside the Soxhlet extractor's thimble; 250 mL of absolute ethanol was added as a solvent to a round-bottom flask that was attached to an isomantle-mounted Soxhlet extractor and condenser; the crushed plant material was loaded into the Soxhlet extractor's thimble. The side arm had glass wool lagged on it. Using the isomantle, which is used to evaporate, the solvent was heated before passing through the apparatus and into the condenser. Put another way, the condenser unit of the apparatus contains the evaporated ethanol, and the Soxhlet extraction procedure heats the solvent (ethanol) to boiling temperature (>78°C). After that, the condensate is poured into the reservoir holding the thimble. The cycle restarts when the solvent level hits the syphon and flows back into the flask. For 3 hours, the dye-stuff extraction was refluxed. In order to acquire the crude solid dyestuff extracts using a water bath, the extract phases produced by the various extraction operations were first distilled to recover some of the solvent and then allowed to evaporate to dryness. To create pure dye samples for the physical and chemical analyses, the crude dye stuff extracts were further air-dried at room temperature [32]. Water and acetic acid are used as solvents to remove the colour. In order to remove the onion peels, 0.5 g of them was extracted, and 45 ml of distilled water was mixed with 5 ml of concentrated acetic acid. As a result, a 1:19 ratio of acetic acid to water was used. A magnetic stirrer was used to continuously swirl the mixture. Throughout the whole experimental time, the procedure was constantly observed. Five millilitre samples of the combination were taken at regular intervals. A 4 millilitre sample was extracted for gravimetric examination. The material was cut up into 1 millilitre for ultraviolet (UV) examination. Every experiment was carried out in batches [33]. The common yellow onion, *Allium cepa*, is used to make Shepherd Textiles Onion Peel Extract, as shown in (**Figure 2**). The dark colours of onion skins are caused by a variety of anthocyanin chemicals, which can be boiled away to create a potent fabric dye. Moreover, onion skins contain a lot of natural tannins, which make them perfect for dyeing cellulose materials like linen and cotton. For pleasing mustard tones, use them at 5% weight-of-fabric on unmordanted fibre. For exceptionally

Figure 2.
Yellow onion skin extract.

Figure 3.
Red onion skin extract.

deep, rich, golden oranges, dye again at 5% weight-of-fabric and mordant with alum. Approved by GOTS (Global Organic Textile Standard) and appropriate for use on fibres that will bear the "organic" label [34].

Yellow onion skin dye imparts golden/yellow, orange, and brown tones. When considering alternative natural dyes, onion skin dye is comparatively colourfast. It works rather well to dye animal-based fibres like silk and wool, but less so on cotton and linen [31, 35]. The colour of red onion skins is deeper than that of yellow, which can be either peachy yellow or even orange, as shown in (**Figure 3**). Thus, make sure to experiment with several onion varieties if you are going for a particular colour [36, 37]. The onion skins were allowed to dry in the sun before being ground into tiny pieces using a grinding machine. After removing the leftovers with a fine sieve, the weight was recorded. Following the processes of drying, crushing, and straining, 805 grammes of skins weighing 1 kg was discovered. **Figure 2** displays the crushed and dried skins. Using an aqueous extraction method, the colour component was removed from the skins. A set amount of crushed skins was extracted at pH 5 using a material-to-liquor ratio of 1:10 (crushed skin weight in grammes; water volume in millilitres) at a boil for 60 minutes. The mixture was chilled after each extraction step, and the dye extracts were then precisely filtered [38].

3. Dyeing process

Dyeing process is based on different steps that are described here.

3.1 Preparing fabrics for dyeing

At pH 9, an aqueous soap solution (15 g/L soap) was used to degum the raw silk fabric. Throughout the treatment, the material-to-liquor ratio was kept constant at 1:50. For 60 minutes, the temperature was gradually increased to 80°C. After degumming, the cloth was cleaned for 10 minutes at 65°C using 2 g/L of detergent. After that, the cloth was treated for 60 minutes at 60°C with a 35% hydrogen peroxide (H_2O_2) (3 mL/L) solution, keeping a material-to-liquor ratio of 1:50 at pH 9 and 60°C.

After that, it was washed for 10 minutes at 65°C using 2 g/L detergent [38]. Wool and nylon textiles were placed in a bath with a material-to-liquor ratio of 1:50, 2 g/l nonionic detergent, and 5 g/l sodium sulphate. After that, the bath's overall temperature was raised to 45 degrees Celsius, and it stayed there for 30 minutes. After rinsing with warm water, wool and nylon were allowed to air-dry. A bath of 5 g/l nonionic detergent and 2 g/L sodium carbonate (Na_2CO_3) was added to the cotton fabric at a material-to-liquor ratio of 1:50. After that, the whole bath was brought to a boil and kept there for an hour. After that, it was rinsed and allowed to air-dry [36].

3.2 Bio-mordants in natural dyeing

Numerous advancements in textile material colouring have been developed, particularly in recent times. The application of natural dyes derived from plants in dyeing procedures is the first of these advances. Historically, natural dyeing techniques have made extensive use of metal salts, often known as mordants. The word "mordant" comes from the Latin word "mordere", which means "to bite". Something that can be attached to a fibre and reacts chemically with natural colourants is called a mordant. Natural dyes fall into two categories: substantive dyes and nonsubstantive dyes, depending on how they are utilised. It is not necessary to pretreat the fabric before applying the substantial dyes (such as turmeric, indigo, and orchil). On the other hand, nonsubstantive dyes (such as fustic, cochineal, madder/alizarin, and logwood) can only colour or work with materials that have previously been mordanted if a mordant is added to the dyebath [15].

The widely recognised vat dyeing technique was used to dye the scrubbed cotton garments. Each batch, which consisted of six pieces of scrubbed cotton cloth, was dyed in turn. Cupric sulphate, ferrous sulphate, potassium dichromate, alum, lemon juice, and aloe vera were used as mordanters for the six cloth samples. One batch of cotton was utilised for each of the three distinct mordanting procedures, such as pre-mordanting, simultaneous mordanting, and post-mordanting, which were performed. Another piece of cloth was likewise dyed without a mordant [39]. The materials need to be dried after being scrubbed and bleached. After that, it is taken to be mordant. Both synthetic and natural mordants were utilised for mordanting. Iron water, myrobalan fruit, and eucalyptus bark were utilised as natural mordants. Alum and ferrous sulphate were utilised as synthetic mordants [14, 40].

Using 5% (on weight-of-fabric) of each of the following mordants: ferrous sulphate, alum, stannous chloride, tannic acid, and tartaric acid; four distinct combinations of mordants were also used: ferrous sulphate-alum (2.5 + 2.5%), ferrous sulphate-alum-tin (2 + 2 + 1%), alum-tin (2.5 + 2.5%), and alum-tin-tannic acid (2 + 2 + 1%) at 60 C for 60 minutes while maintaining a material-to-liquor ratio of 1:50. Once more, alum and stannous chloride were utilised independently, with cream of tartar (CT) acting as a mordant helper [41].

3.3 Fastness of the onion skin dye on fabrics

Regarding the colour yellow for 3 days, 500 grammes of linden, 1 kg of onion skin, and the equal quantity of sage plants are stored in cold water [42]. A vivid chromatic colour known as lemon yellow is created when wool that has been mordanted with alum is cooked in the same water for an hour. Tones that are coppery yellow can be created by boiling wool that has been mordanted with chrome in this water [31]. The fastness of onion skin, which is generally considered to be low, varies based on the fastness of the other dyestuffs used in combination with it. Dried quince leaves are coloured red by soaking them in water for a week along with red onion skins [43]. As much as 500–700 grammes of wool is needed for every kilogramme of pre-mordanted (alum) wool in order to produce 1 kilogramme (kg) of the material. An hour of boiling produces a brilliant peach red colour [44].

3.4 Examples of successful natural dye projects using onion skin

Onion peels are a rich source of plant components known as flavonoids, and they can also be used to make a dye called pelargonidin. Additionally, a suitable piece of leather was stained using pelargonidin dye, which was produced from onion peels using acetic acid. In light of this, the current study examines the practical application of onion peels for advantageous purposes, such as the extraction of natural dyes, as a feasible method of developing environmentally safe goods from agricultural waste [33]. Excellent ultraviolet protection factors (UPFs) (50+) can be achieved when dyeing fabrics with or without mordants, with the exception of nylon, which has a low UPF when dyed with onion. It is advised to avoid skin cancer by using these naturally coloured materials, and the textile colouring business is encouraged to use their colourants [36]. Using a dyeing process based on natural components taken from the outer shell colourant (quercetin) of onions using water and organic solvents, coloured silk fabric was created for textiles [45].

4. Conclusion

In this chapter, we concluded that onions (*Allium cepa*) are of different types and their skin (peel) is not wasted but used as dyes due to having different phytochemicals, quercetin, pelargonidin, etc. Onion skin gives different colours with and without different mordants. These dyes are used for not only fibres but also leather and other substances. Excellent UPFs (50+) can be achieved when dyeing fabrics with or without mordants, with the exception of nylon, which has a low UPF when dyed with onion. It is advised to avoid skin cancer by using these naturally coloured materials, and the textile colouring business is encouraged to use their colourants.

Author details

Sidra Mubeen[1*], Tahira Batool[1] and Ghulam Sabir Hussain[2]

1 Department of Chemistry, The Women University Multan, Punjab, Pakistan

2 Department of Agronomy, Bahauddin Zakariya University Multan, Punjab, Pakistan

*Address all correspondence to: sidra.mubeen@wum.edu.pk

References

[1] Yadav S et al. A brief review on natural dyes, pigments: Recent advances and future perspectives. Results in Chemistry. 2023;**5**:100733

[2] Gürses A et al. Classification of dye and pigments. Dyes and Pigments. Vol. 1. Springer; 2016. pp. 31-45

[3] Brunello F. The Art of Dyeing in the History of Mankind. Neri Pozza Aatcc; 1973

[4] Abel A. The history of dyes and pigments: From natural dyes to high performance pigments. In: Colour Design. Woodhead Publishing Series in Textiles Elsevier; 2012. pp. 557-587

[5] Bishal A, et al. Natural Dyes: Its Origin, Categories and Application on Textile Fabrics in Brief. European Chemical Bulletin. 2023;**12**(8):9780-9802

[6] Prabhu K, Bhute AS. Plant based natural dyes and mordants: A review. Journal of Natural Product and Plant Resources. 2012;**2**(6):649-664

[7] Ardila-Leal LD et al. A brief history of colour, the environmental impact of synthetic dyes and removal by using laccases. Molecules. 2021;**26**(13):3813

[8] Yusuf M, Shabbir M, Mohammad F. Natural colorants: Historical, processing and sustainable prospects. Natural Products and Bioprospecting. 2017;**7**:123-145

[9] Saxena S, Raja A. Natural dyes: Sources, chemistry, application and sustainability issues. In: Roadmap to Sustainable Textiles and Clothing: Eco-Friendly Raw Materials, Technologies, and Processing Methods. Singapore: Springer; 2014. pp. 37-80

[10] Das A et al. An innovative cell imaging by beet root extracted pigment. SpectrochimicaActa Part A: Molecular and Biomolecular Spectroscopy. 2020;**230**:118037

[11] Vuthiganond N et al. Ecological dyeing of acrylic yarn with colorant derived from natural lac dye. Journal of Metals, Materials and Minerals. 2022;**32**(4):71-78

[12] Islam M, Mostafa M. Textile dyeing effluents and environment concerns-a review. Journal of Environmental Science and Natural Resources. 2018;**11**(1-2): 131-144

[13] Pizzicato B et al. Advancements in sustainable natural dyes for textile applications: A review. Molecules. 2023;**28**(16):5954

[14] Iqbal S, Ansari TN. Extraction and application of natural dyes. Sustainable Practices in the Textile Industry. 2021;**1**:1-40

[15] Pinheiro L et al. Biomordants and new alternatives to the sustainable natural fiber dyeings. SN Applied Sciences. 2019;**1**(11):1356

[16] İşmal ÖE, Yıldırım L. Metal mordants and biomordants. In: The Impact and Prospects of Green Chemistry for Textile Technology. Woodhead Publishing in Textile Series Elsevier; 2019. pp. 57-82

[17] Singh AP, Kumar S. Applications of tannins in industry. Tannins-Structural Properties, Biological Properties and Current Knowledge. 2020;**1**:1-13

[18] Lawande K. Onion. In: Handbook of Herbs and Spices. Woodhead Publishing

Series in Food Science, Technology and Nutrition Elsevier; 2012. pp. 417-429

[19] Griffiths G et al. Onions—A global benefit to health. Phytotherapy Research: An International Journal Devoted to Pharmacological and Toxicological Evaluation of Natural Product Derivatives. 2002;**16**(7):603-615

[20] Griffith L, Griffith F. Onions, Onions, Onions: Delicious Recipes for the World's Favorite Secret Ingredient. Vol. 1. Houghton Mifflin Harcourt; 2002

[21] Teshika JD et al. Traditional and modern uses of onion bulb (Allium cepa L.): A systematic review. Critical Reviews in Food Science and Nutrition. 2019;**59**(sup1):S39-S70

[22] Benítez V, et al., Characterization of industrial onion wastes (*Allium cepa* L.): Dietary fibre and bioactive compounds. Plant Foods for Human Nutrition. 2011;**66**:48-57

[23] Ng A, Smith A, Waldron K. Effect of tissue type and variety on cell wall chemistry of onion (Allium cepa L.). Food Chemistry. 1998;**63**(1):17-24

[24] Sharma K et al. Economical and environmentally-friendly approaches for usage of onion (Allium cepa L.) waste. Food & Function. 2016;**7**(8):3354-3369

[25] Stoica F et al. A comprehensive review on bioactive compounds, health benefits, and potential food applications of onion (Allium cepa L.) skin waste. Trends in Food Science & Technology. 2023;**141**:104173

[26] Sagar NA et al. Physicochemical and thermal characteristics of onion skin from fifteen Indian cultivars for possible food applications. Journal of Food Quality. 2021;**2021**:1-11

[27] Bello M et al. Characterization of domestic onion wastes and bulb (Allium cepa L.): Fatty acids and metal contents. International Food Research Journal. 2013;**20**(5):2153

[28] Kumar M et al. Onion (Allium cepa L.) peels: A review on bioactive compounds and biomedical activities. Biomedicine & Pharmacotherapy. 2022;**146**:112498

[29] dos Santos C, Antimicrobial potential of Satureja montanaagroindustrial byproducts as possible ingredients in animal feed in Jéssica Daniela. 2018. pp. 1-101 core.ac.uk

[30] Campone L et al. Response surface methodology to optimize supercritical carbon dioxide/co-solvent extraction of brown onion skin by-product as source of nutraceutical compounds. Food Chemistry. 2018;**269**:495-502

[31] Deveoglu O. A Review on Onion Skin, a Natural Dye Source. Journal of Textiles, Coloration and Polymer Science. 2022;**19**(2):307-319

[32] Egbujor MC, Agunwa CI, and Chidebelu IC. Extraction and Characterization of Natural Dyestuffs from Onion (*Allium Cepa*) Bulb, Carrot (*Daucuscarotal.*) Root and Turmeric (*Curcuma Longa*) Root. Iconic Research and Engineering (IRE) Journals. India. 2023;**6**(8):130-141

[33] Mohan R et al. Studies on natural dye (Pelargonidin) extraction from onion peel and application in dyeing of leather. International Journal of Recent Engineering Science. 2020;**7**(1):14-22

[34] Wadhwa M, Bakshi M, Makkar H. Wastes to worth: Value added products from fruit and vegetable wastes. CABI Reviews. 2016;**2015**:1-25

[35] Hamdy DM, Hassabo AG. Various natural dyes from different sources. Journal of Textiles, Coloration and Polymer Science. 2021;**18**(2):171-190

[36] Gawish S et al. UV protection properties of cotton, wool, silk and nylon fabrics dyed with red onion peel, madder and chamomile extracts. Journal of Textile Science & Engineering. 2016;**6**(4):1-13

[37] Pucciarini L et al. Onion (Allium cepa L.) skin: A rich resource of biomolecules for the sustainable production of colored biofunctional textiles. Molecules. 2019;**24**(3):634

[38] Uddin MG. Effects of different mordants on silk fabric dyed with onion outer skin extracts. Journal of Textiles. 2014;**2014**:405626

[39] Zubairu A, Mshelia YM. Effects of selected mordants on the application of natural dye from onion skin (Allium cepa). Science and Technology. 2015;**5**(2):26-32

[40] Samanta AK, Konar A. Dyeing of textiles with natural dyes. Natural Dyes. 2011;**3**(30-56):212-222

[41] Dutta P et al. Impacts of natural and synthetic mordants on cotton knit fabric dyed with natural dye from onion skin in perspective of eco-friendly textile process. Materials Today: Proceedings. 2021;**47**:2633-2640

[42] Patil BS, Pike LM. 444 PB 253 distribution of quercetin and quercetin glycosides in different-colored varieties and genotypes of onion (Allium cepa L.) tissues. HortScience. 1994;**29**(5):494-495

[43] Mercado SDLC. A Social Marketing Intervention Model to Influence the Social Acceptance of Natural Dyes, from Domestic Food Waste Resources in Mexico without Metallic Chemical Additives for Textile Dyeing. University of Huddersfield; 2021

[44] Deveoglu O. Journal of Textiles, Coloration and Polymer Science. 2022

[45] Miah MR et al. Comparative analysis of colour strength and fastness properties on extracts natural dye from onion's outer shell and its use in eco-friendly dyeing of silk fabric. International Journal of Photochemistry and Photobiology. 2017;**2**(1):1-8

Section 2

Synthetic Dye and Applications

Chapter 5

Azobenzenes: Photoswitching and Their Chemical Sensor Application

Girish Chandra, Ujala Rani, Birkishore Mahto and Gopal Kumar Mahato

Abstract

Azobenzene is a well-known dye that undergoes fast *trans-cis* photoisomerization and has been widely studied and used in the development of organic functional materials. Due to its selective isomeric distribution in the excited state, azobenzene has been used as a photoswitch in the storage of information on a molecular level, photo-controllable catalysis, solar light harvesting, photo-pharmacology, optical-to-mechanical energy conversion, molecular electronic, and photonic devices. Furthermore, the characteristic and distinguishable photoelectronic properties of *trans* and *cis* azobenzene have been recently used in the sensing properties of different ions and the recognition of molecules. Here, we are going to review the recent literature where different intermolecular forces show the supramolecular properties under the stimuli of photo-light.

Keywords: azobenzene, photoswitching, photoisomerization, chemical sensors, molecular recognition

1. Introduction

Azobenzene is one of the old known chemicals among few, with which people are fascinated due to its impressive color and have been using it rigorously as a dye. Over the last two centuries, we have witnessed a tremendous change in the application pattern of azobenzene, initially known as a dye, but now evolved as a superpowered tiny engine that could do varieties of works [1]. Thus, recently azobenzene has shown tremendous utility in every day of our life. The design of materials is used for many purposes, *viz*, solar light harvesting, optical-to-mechanical energy conversion, molecular electronic and photonic devices, in medicinal chemistry, photo-pharmacology, drug delivery, etc. The presence of two benzene rings connected by -N=N- bond makes them have a special photophysical property and has been widely used for light-responsive materials. Many probes have been developed as chemical sensors for the detection of varieties of ions and molecules [2, 3].

Furthermore, azobenzenes have been used as photoswitch materials due to their light-responsive reversible *trans* $\leftrightarrows$ *cis* photoisomerization. The change in the configuration of azobenzene opens up a new avenue to explore its physical, chemical, biological, and material properties, since it causes changes in geometries, conformations, and assemblies of molecules [4, 5]. Molecular switches enable the storage of information on a molecular level and since the difference of energy is approximately 47 KJ.mol^{-1}, modulation of the above-mentioned properties could thus be easily achieved reversibly, according to our choice by using noninvasive and versatile stimuli such as light. Azobenzene undergoes symmetry-allowed excitation to π-π^* ($S_0 \rightarrow S_2$) transitions, followed by rapid relaxation to the S_1 state. Then, there is isomerization of *trans*-azobenzene along with movement from the Frank-Condon region and $S_1 \rightarrow S_0$ relaxation. Over the process, there is a decrease in the intensity of the π-π^* band with a blue shift and an increase in the intensity of the symmetry-forbidden n-π^* band ($S_0 \rightarrow S_1$). The *trans* → *cis* photoisomerization leads to nonplanar form with the increase in the dipole moment of *cis* form. The reverse transformation from *cis* to *trans* proceeds *via* either the photochemical n-π^* transition or the thermal process. Depending on the substituents present in the benzene rings, the kinetics of the photoisomerization reaction is determined. Both fast and slow reactions are useful for different applications. Photoswitching properties of azobenzene have been studied for decades, but the last few decades have made this photo-intertransformation, a hot research area, since by appending suitable substituents on azobenzene made it a smart material [6, 7]. Thus, wide applications in energy harvesting, sensing [8], electronics [9], biology [10], soft robotic systems [11], etc., have been explored.

Due to wide application in diagnostics, imaging, and environmental protection concerns, the development of receptors for selective recognition of anions is one of the hot topics in modern science and many colorimetric and fluorescent materials have been developed [12–14]. There is growing interest to develop probes for ionic or molecular recognition, which are controlled by external stimuli like light. Light changes the conformation or configuration of the probes, and thus binding affinities could be controlled. In the last two decades, many photoswitchable probes, *viz*, azobenzene [15], spiropyran [16], diarylethene [17], etc., have been developed, which we found to have many applications in materials science and biology. The azobenzene photoswitch is particularly interesting where a planar *trans* form changes to a nonplanar *cis* form, and during these interconversions, there is a dramatic change in the dipole moment of the compounds (**Figure 1**). Due to easy synthesis, robustness during prolonged exposure to light, and easy functionalization, azobenzenes are most extensively used in the development of materials.

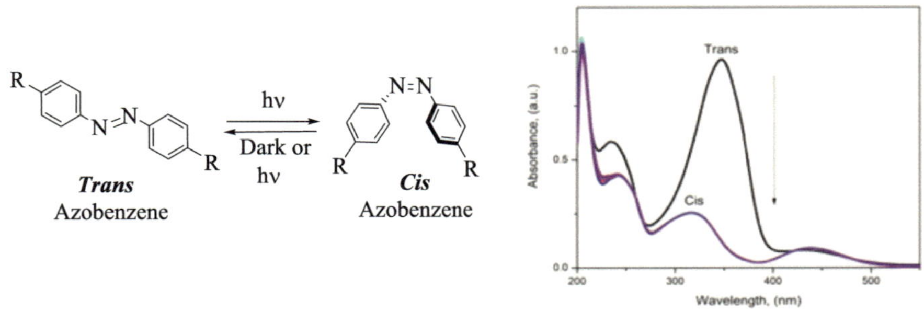

Figure 1.
Structure of trans and cis azobenzene and their interconversion initiated by photoirradiation. Characteristic UV-visible absorbance of trans (black colored) and cis azobenzene (blue colored). Reprinted (adapted) with permission from Ref. [4]. Copyright © 2024, Wiley.

2. Azobenzene photoswitching in sensing application

Jiang et al. reported the synthesis of an azobenzene-containing phenyl-1,2,3-triazole motif photoswitchable foldamer **1**, which showed a controllable change in binding affinity for anions. Thus, compound xx upon photoirradiation with 339 nm shows an 80% descent in the π-π^* transition and increment in the n-π^* transition and irradiation at the visible light, band of the π-π^* transition reverts back. Both *trans* and *cis*-configurations of compound **1** were confirmed by two-dimensional nuclear magnetic resonance (2D NMR) experiments. The addition of chloride ion (Cl-) from tetrabutylammonium chloride (Bu_4NCl) in the ground state suggested an interaction with the inner cavity of the foldamer that participates in hydrogen bonding, as suggested by the proton nuclear magnetic resonance (^{1}H-NMR) analysis. Furthermore, titration with other ions also showed a similar kind of interaction (1:1), and their binding constant was found to be 70 M^{-1} for Cl^-, 22 M^{-1} for Br^-, 38 M^{-1} for HSO_4^-, 11 M^{-1} for I^-, and 10 M^{-1} for NO_3^- ions. The *cis* configuration dramatically enhances the interaction and here the presence of a triazole ring has more significant contributions, as suggested by the ^{1}H-NMR. Thus, binding constants were found to increase up by 290 M^{-1} for Cl^-, 87 M^{-1} for Br^-, 66 M^{-1} for HSO_4^-, 31 M^{-1} for I^-, and 21 M^{-1} for NO_3^- ions. Thus, binding affinities are highly dependent on the size and shape of anions (**Figure 2**) [18].

Bandyopadhyay et al. recently developed a macrocyclic azobenzene-based fluorescent receptor 2 that very selectively recognizes ATP (adenosine triphosphate) in the 2-*trans* form. There was a sevenfold increment of fluorescence emission observed when 2-*trans* interacted with ATP (10 equivalent), on the other hand, ADP (adenosine diphosphate) and GDP (guanosine diphosphate) showed an increment only 3.5 times only. But upon irradiation by 366 nm, it causes photoisomerization to 2-*cis* to show altered selectivity to GTP (guanosine triphosphate). Interestingly, 2-*trans* exists in the protonated form through the pyridinium nitrogen atom, even in the neutral condition. Receptor 2-*trans* showed strong affinity with ATP^{4-} due to high Coulomb interaction between the negatively charged oxygen atom of the anions and the positively charged protonated form of the receptor. Thus, due to different binding affinities of various phosphorylated species in their *trans* and *cis* form, system 2 was used for monitoring enzymatic phosphorylation reaction with tyrosine kinase enzymes (**Figure 3**) [19].

The rate of interconversion between *trans* and *cis* configurations depends upon many factors. The half-life of a few *cis* -compounds lasts up to weeks but the half-life of some compound could instantly revert back. Both fast and slow individual interconversion processes have been utilized for many purposes. The change in pH

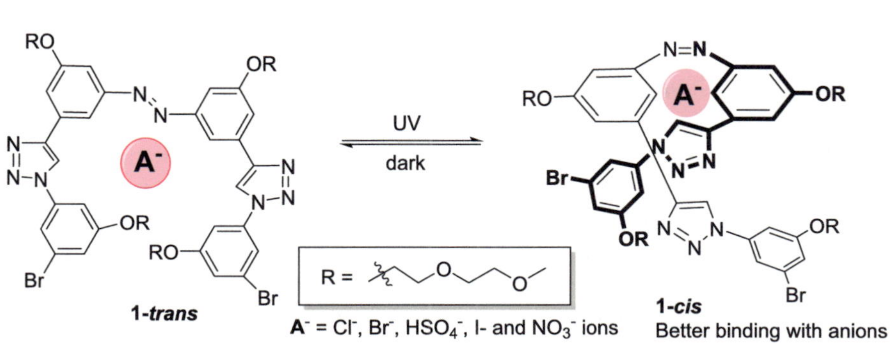

Figure 2.
Photoswitch 1 and a different binding model proposed for its two photoisomers to anions.

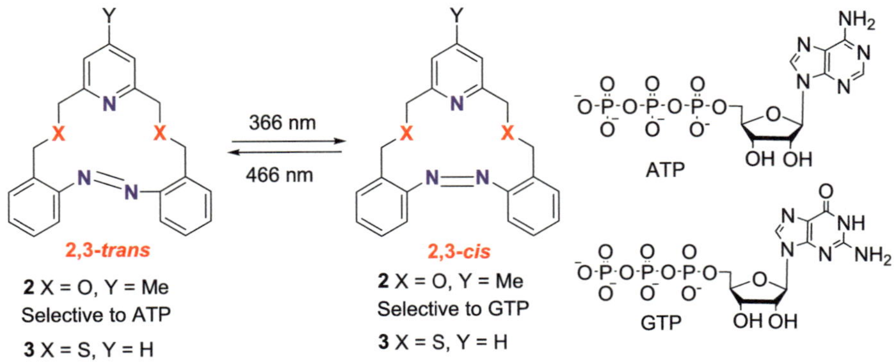

Figure 3.
Photomodulation of the receptors 2 and 3.

modulates the stability and the presence of substituent at the *ortho*-position increases the stability of *cis* –isomer [20, 21]. Furthermore, the type of solvent (polar or nonpolar) and intermolecular H-bonding also affect the stability of the *cis* isomer. Bandyopadhyay *et al.* reported the synthesis of macrocyclic azobenzenes (2 and 3, Scheme 2) that underwent exclusive photo-transformation to *cis* -form. The addition of Cu^{2+} and continuous exposure to ultraviolet (UV) radiation make the *cis* compound stable. But, interestingly, on ceasing of UV light, the *cis* compound quickly gets converted back to the *trans* form. The Cu^{2+} ions also bind to the initial *trans* compounds with the binding constant of 9.9 × 102 M^{-1}, as determined by ^{1}H NMR. Irradiation with UV light to this Cu^{2+} complex led to slow photoisomerization to *cis* form [22]. However, on ceasing of the UV light source, the *cis* form immediately goes back to the initial *trans*-Cu^{2+} complex within 85 minutes with first-order rate constant of 1.75×10^{-5} min^{-1}, unlike the original *cis* complex ($t_{1/2}$ = 92 days at 298 K). The Ag + ion was also found to exhibit the same behavior but slower than the Cu^{2+} ion.

The concept of photoswitching has been widely used in biological gated systems to timely open and close the ion channels in a controlled fashion for regulating vital functions of nerve *transmission*, osmotic regulation, ion homeostasis, muscle excitation, growth, and development. Natural stimuli, such as variations in pH, light, chemical messengers, and membrane potential differences, are what cause the physiological systems' aforementioned regulating activity. So, misregulation in the above stimuli causes numerous diseases, collectively called channelopathies. Recently, a significant effort has been made to mimic the natural ion *transport* systems. And in this direction, the azobenzene-based photoswitch has proved to play a significant role. Thus, binding with ions in a supramolecular concept to use twisted *cis* compound as an anion carrier has been widely used. Yet, intramolecular hydrogen-bound *trans*-azobenzene diamide system 4, which was recently developed and synthesized by Talukdar et al., provides a thermally stable inactive state in the *cis* -conformation and creates an effective photoswitchable chloride carrier [23]. Thus, in compound 4-*trans*, the amide N-H_1 and C_{Ar}-H_2 create the anion-binding site. In the lipid bilayer membrane, receptor 4 was found to facilitate the transport of Cl^- through the formation of a sandwich complex with a chloride ion. Nevertheless, poor anion binding and transport resulted from a change in the planarity and closeness of the anion-recognizing groups caused by the photoisomerization of receptor 4 at 365 nm. Different substituents were taken based on lipophilicity, as per the Lipinski rule and it was found that 4-(trifluoromethyl) phenyl aminoformyl and a propionamide groups provided the best half-maximal effective concentration (EC_{50}) = 0.199 μM and Hill coefficient of $n \sim 2$ when measured across EYPC-LUVs⊃HPTS (egg yolk phosphatidylcholine-large unilamellar vesicles-pyranine (8-hydroxypyrene-1,3,6-trisulfonic acid)) (**Figure 4**).

Bandyopadhyay et al. demonstrated a photoswitchable assembly-disassembly process through the photoisomerization of *trans* and *cis* interconversion. Thus, azobenzene compound 5 was synthesized, which has a suitable binding pocket that binds to Al^{3+} ions that set up a linear polymeric assembly with the *trans* form of photoswitch. But photoirradiation of the above-mentioned complex generated a *cis* form where intramolecular chelation disrupts the assembly. However, the coordination assembly regenerated when the *trans* complex was irradiated with visible light [24]. The development of assembly was the characteristic property of the Al^{3+} complex (confirmed by dynamic light scattering (DLS) showing the average hydrodynamic diameter, d of 448 nm.) since the original azobenzene did not show any aggregation behavior. The binding constant for the association of Al^{3+} ions with *trans*-5 was calculated to be 162 M^{-1} (from ^{1}H NMR) with 1:1 interaction (from Jobs Plot). Furthermore, investigation by using ^{1}H NMR confirms that terminal N-atoms participated with the Al^{3+} ion (**Figure 5**).

Kalow *et al.* developed and used a series of azobenzene boronic acids for reversible diol binding *via* photochemical isomerization reaction. It was observed that the *trans*-azobenzene having *ortho*-boronic acid substituent weakly binds with diols to form boronic esters, but the isomerization of the azobenzene to its *cis* isomer enhances diol binding. Furthermore, a correlation between diol binding and the photostationary state was investigated, which demonstrated that change in the wavelengths of irradiation yielded different quantities of the bound diol. This pattern was observed with most of the diols including cyclic diols, nitrocatechol, biologically relevant compounds, and polyols. The thermodynamics of esterification was explained by competition experiments and computational studies which suggested that the changes in the binding affinity originate from the stabilization by intramolecular H-bonding interactions between the boronic acid O-H and proximal heteroatoms, and the destabilization of the *trans*-azobenzene by steric interactions (**Figure 6**). Authors demonstrated the application of the above photoswitchable boronic acid-ester equilibria to a dynamic covalent hydrogel made by

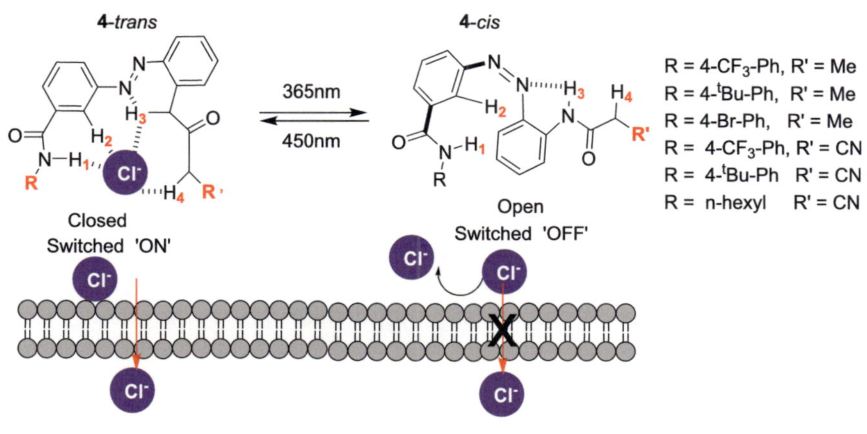

Figure 4.
Design and working principle of light regulatory synthetic ion transporter. Chemical structures of receptors trans-4 and cis-4.

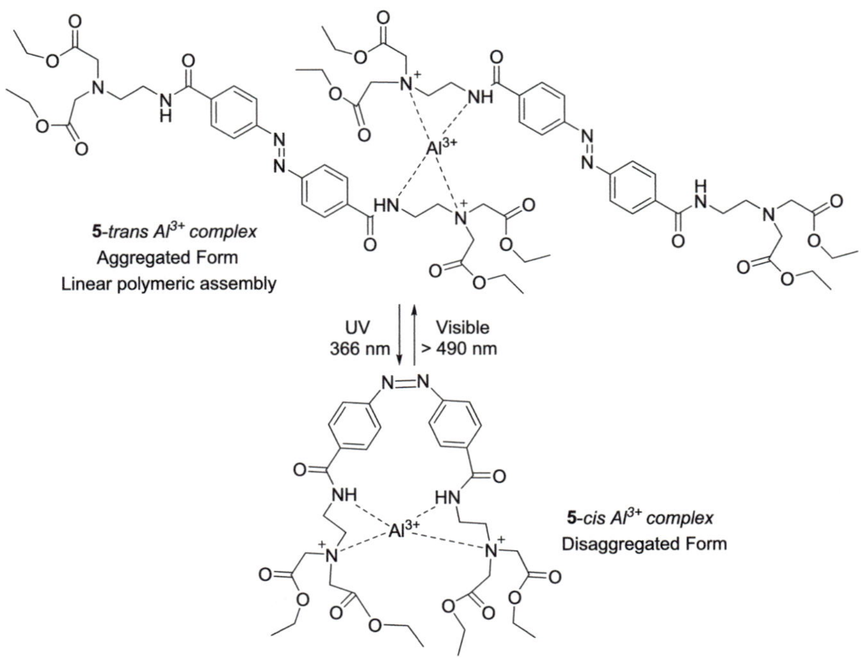

Figure 5.
Putative complexation mode of the Al^{3+} ion with the trans and cis forms of 5.

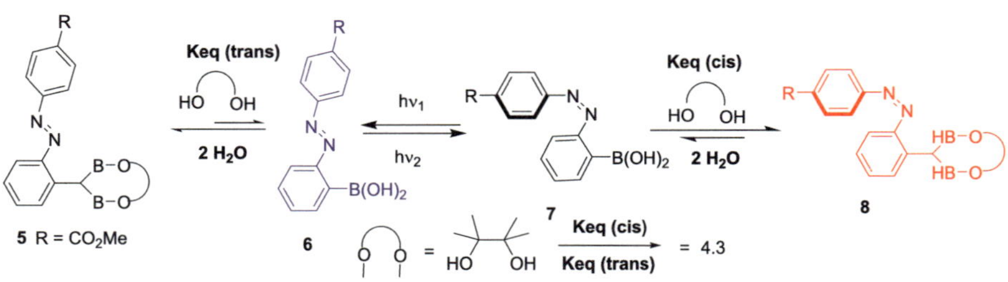

Figure 6.
The equilibrium that results from photoisomerization of the photoswitch between ortho-substituted azobenzene boronic acids and diols, where the cis isomer has a greater binding affinity with diols than the trans isomer. The cis binding constant is 4 times larger than the trans binding constant when the diol is pinacol.

diol-substituted amine-terminated 4-arm poly(ethylene glycol) (PEG) and found that boronic esters can be stiffened with red light [25].

Langaton *et al.* synthesized *ortho*-chloro azobenzene 9, with electron-deficient iodo-triazole motifs, the first halogen bonding photoswitchable anion receptors, wherein it was demonstrated that chloride binding may be reversibly modulated by the red and blue light. The *trans*-form of compound 9, which binds weakly with Cl-, appeared to show a 50-fold enhancement in chloride-binding affinity observed for the *cis* -isomer. The irradiation of 9 affords long *cis* isomer lifetimes (half-life of ~1 week), which was advantageous for switching binding applications. Furthermore, dicationic iodotriazolium system 1-Me-BarF (), *trans*-10 was synthesized by methylation and subsequent anion exchange and it was found to serve as a more promising host for chloride and was enhanced relative to *trans*-9. But, the comparatively switchable anion affinity in *cis* was maintained but diminished albeit significantly ($K_i^{trans} = 349\ M^{-1}$, $K_i^{cis} = 945\ M^{-1}$,

K_{cis}/K_{trans} = 2.7, DMSO (dimethyl sulfoxide)-d_6). This may be argued due to charge repulsion in the *cis* -isomer where both binding sites are in close proximity. System 10 showed unprecedented photo-regulated halogen bonding mediated halide abstraction catalysis for a Friedel-Crafts alkylation and Mukaiyama aldol reaction. Here also, the stronger binding *cis* isomers acted as the more potent catalytic species (**Figure 7**) [26].

Dąbrowa *et al.* synthesized a new tetra-*meta*-substituted azobenzene *trans*-11-based receptor for oxoanions, in particular, tetrahedral $H_2AsO_4^-$ and $H_2PO_4^-$. The photoirradiation of compound *trans*-11 led to *cis* – 11, which was evaluated by ^1H NMR spectroscopy. Titration with anions revealed that nonplanar *cis* –11 was a better host for anions than *trans*-11, which was argued from the greater ability of four amide NH protons in the *cis* -state to cooperatively bind with the oxyanions. Furthermore, the planar *trans*-11 was found to show poorer binding affinities, since its binding groups are too distant from each other and cannot cooperatively bind anions. Interestingly, *meta*-substituted in compound *cis* –11 was shown to significantly increase the thermal stability and very resistance of the *cis* -state to rapid decay triggered by anion binding (**Figure 8**) [27].

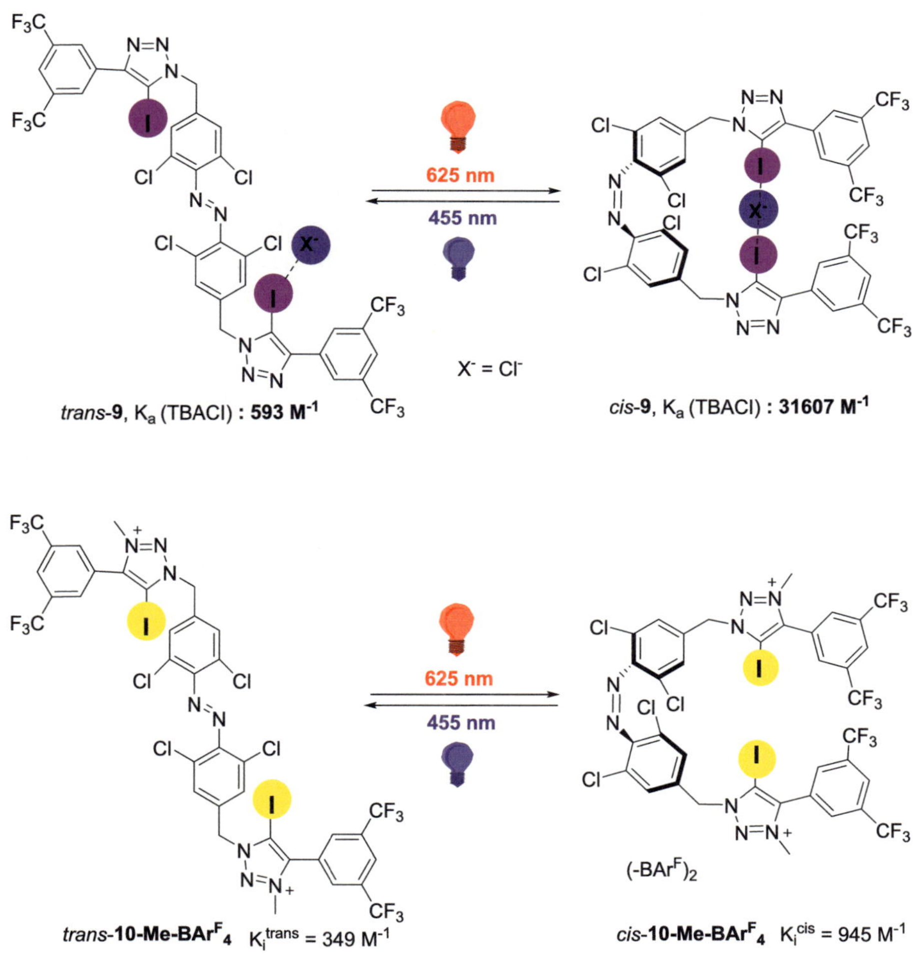

Figure 7.
Photoswitchable anion receptors 9 and 10.

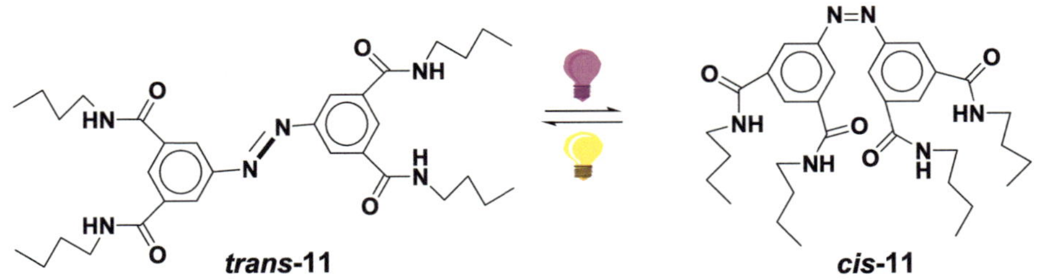

Figure 8.
Photoisomerization of receptor 11.

Jarosz *et al.* developed an efficient cesium carbonate (Cs_2CO_3)-templated macrocyclization strategy to synthesize chiral photoresponsive macrocyclic azobenzene compound 12 with protected sucrose. Green light (530 nm) irradiation causes the photoisomerization to long-lived *cis* –12 (25 days), which undergoes photoswitching back to *trans*-12 by using blue light (410 nm). Both compounds bind with potassium and other alkali *metal* cations but the *cis* –12 exhibits higher binding affinity and selectivity for cation than the *trans*-12 ($K_{cis}/K_{trans} \leq 4.1$). ^{1}H NMR titrations and density functional theory (DFT) calculations confirmed that, in addition to the polyether residues, the sucrose ring oxygen and azobenzene nitrogen atoms also markedly contribute to cation recognition (**Figure 9**) [28].

Flood et al. synthesized a unique bis-azobenzene system *trans-trans*-13 that contains aryl-triazole foldamers and has a helical structure in the ground state. Irradiation of light 365 nm changes the configuration to *cis -trans*-13 and *cis-cis* –13, and changes to random coil structure. Thus, by irradiating the system with 365 and 436 nm light, it is possible to destabilize and stabilize the helix. The probe *trans-trans*-13 was found to show good affinity with the chloride ion having an association energy of $\Delta G = -20$ KJ mol^{-1} ($K_a = 3000$ M^{-1}). When the same receptor solution was exposed to the UV light, the chloride affinity was reduced to $\Delta G = -15$ kJ mol^{-1} ($K_a = 380$ M^{-1}). Furthermore, exposure to 436 nm light restored the dominance of the *trans-trans* isomer in solution and the stronger chloride affinity. Thus, the chloride levels can be regulated with light. In order to verify this, a conductivity experiment was carried out using equimolar concentrations of 13 and tetrabutylammonium

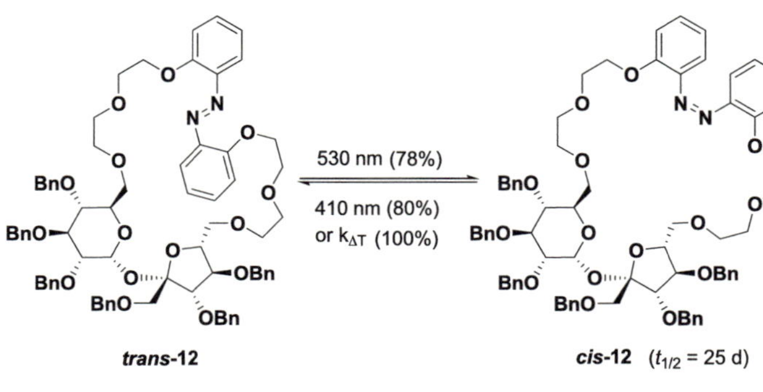

cation	trans	cis	K_{cis}/K_{trans}
Li^+	37	21	0.57
Na^+	210	446	2.12
K^+	1023	2776	2.71
Cs^+	90	370	4.10

Figure 9.
Photoswitching of host 12 and complexes of 12 with 10 equiv. of alkali metal triflates in methyl cyanide (MeCN) (50 μM) at 298.0 ± 0.1 K using green light (LED 530 nm) and blue light (LED 410 nm), respectively. Stability constants K_a (M^{-1}) for complexes of hosts trans-12 and cis –12 with alkali metal cations.

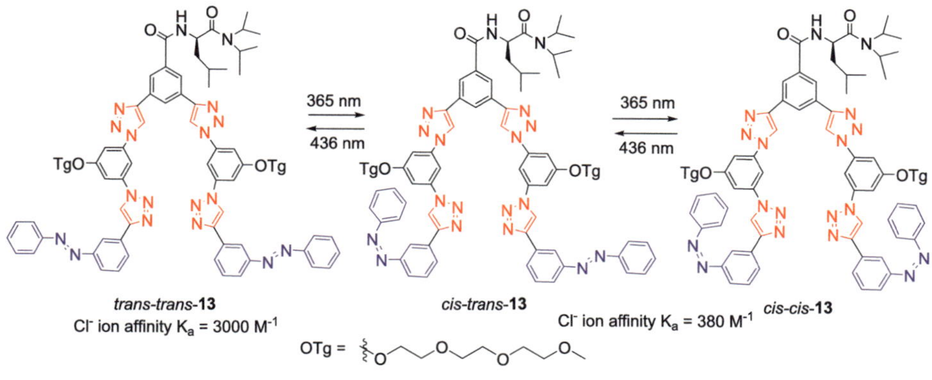

Figure 10.
Representations of the helical-to-random coil equilibria for $13_{trans\text{-}trans}$ and the UV-driven PSS (66% $13_{cis\text{-}trans}$ + 33% $13_{cis\text{-}cis}$).

chloride (TBACl) (1 mM, CH_3CN, 298 K). It was discovered that the visible polystyrene sulfonate (PSS) will have a free chloride concentration of 0.23 mM, which will rise to 0.56 mM when exposed to UV light (**Figure 10**) [29].

Russell *et al.* designed and synthesized a novel (pyridylazo) phenol derivative 8-[4-(2 pyridylazo)phenoxy]octyl disulfide 14 and used this for the formation of a monolayer film on gold-coated optical waveguides through self-assembly. The 14-self-assembled monolayer (SAM) in the *trans* isomeric form provides a bidentate ligand that chelates with *metal* ions, such as Ni^{2+} and Co^{2+}, and results in a strong red-shifted UV-visible absorption at 433 nm. Further photoirradiation with waveguided UV light at 365 nm *via* the evanescent field, the 14-SAM underwent photoisomerization to the *cis* form. But, *cis* – 14-SAM was not able to chelate with *metal* ions. However, the above process was reversible, thus irradiation at visible light at 439 nm, *cis* – 14-SAM converted back to *trans*-14-SAM. Thus, through this process, the *trans*-14-SAM was used to detect Ni^{2+} in the range 9.6×10^{-5}–2.44×10^{-3} M (**Figure 11**) [30].

Jurczak et al. developed a system where the lifetime of *cis* isomer could be decreased by the introduction of anions. Thus, azobenzene urea derivatives 15–16 were synthesized. Photoirradiation of compounds 15 and 16 revealed the decrease in the *cis* lifetime of 16 due to the presence of urea substituent (*cis* –15 = 13.1 h, *cis* –16 = 1.8 h). Interestingly, the addition of different anions decreases the lifetime of the *cis* –16 isomer (with F^- = 15 min, $MeCO_2^-$ = 8 min, $PhCO_2^-$ = 25 min, HSO_4^- = 8 min, and $H_2PO_4^-$ = 28 min). Thus, acetate ions have a high-rate constant for *cis -trans* isomerization in both cases (for *cis* – 15 = 31×10^{-4} s, for *cis* – 16 = 14×10^{-4} s), which was thought to be due to its high basicity and geometry being complementary to the urea group. Thus, varying the nature and concentration of anions can be used for precise control of the thermal *cis* →*trans* isomerization rates. Furthermore, these changes could be easily stopped by the addition of an acid and restored again by the addition of the base (**Figure 12**) [31].

Manabe et al. reported a new synthetic azobenzene ionophore attached to a crown ether ring and ammonium ion tail, 19, which was capable of facilitating intramolecular complexation. Thus, photoirradiation of compound *trans*-19 photoisomerizes to *cis* –19, which was found to show that thermal *cis-trans* isomerization was slower as compared to free amine analog, and the rate was accelerated selectively by the addition of K^+ ion (**Figure 13**) [32].

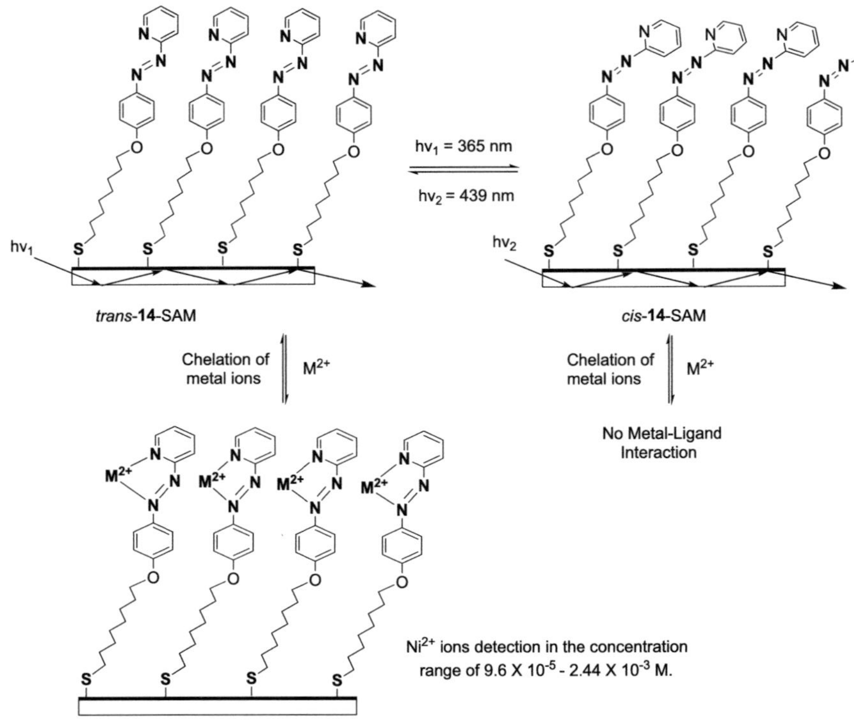

Figure 11.
Diagrammatic summary of the photoswitching and metal-ion chelation of the 14-SAM.

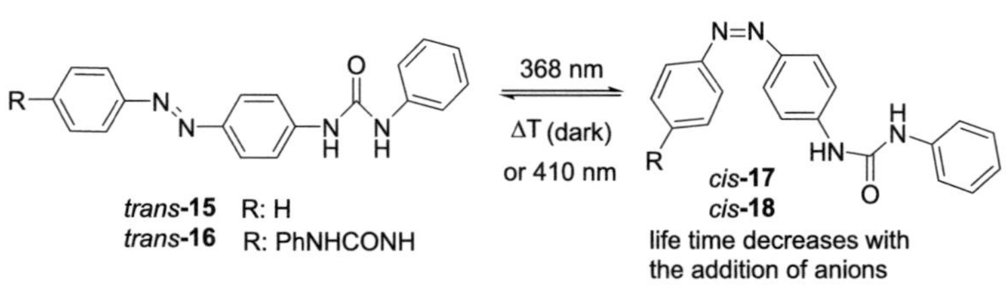

Figure 12.
Trans-cis isomerization of the urea derivative of azobenzenes.

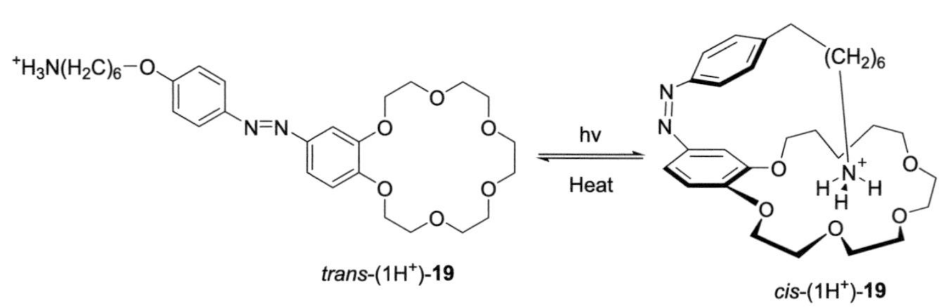

Figure 13.
Trans-cis isomerization of the ammonium ion attached to the azobenzene derivative.

3. Conclusion

This chapter aims to highlight the recent literature on the emerging sensing application through multifunctional azobenzene photoswitching. The colorimetric sensing could be easily detected through the *trans-cis* isomerization due to a change in the inherent UV-vis absorbance pattern of the *trans-cis* compounds. Characteristically, the simple and tiny -N=N- azo functional group could be incorporated into any molecule and have more chances in this category of compound to modulate and get better and better probes for sensing and other applications. This diverse nature including acyclic, cyclic, and macrocyclic azobenzenes has been developed, which have their own characteristic photophysical properties. Also, a good number of works have been done to stabilize the *cis* -isomer by modulating the substituent nearer to the azo functional group. In this chapter, we summarized the latest few and selected examples to highlight the sensing application of photoswitchable azobenzene, which could be easily monitored by UV-visible spectroscopy. The progress in this field is continuing and we expect that we will see a few reports in the near future where fluorescence emission change of azobenzene is used when interacting with ions through photoswitching.

Acknowledgements

The authors, Chandra, Rani, B. Mahto, and Mahato, are grateful to the CUSB-Gaya for providing the research infrastructure. The author Rani also thanks the CUSB-Gaya for the research fellowship from the CUSB and the author Mahto thanks the UGC New Delhi for the research fellowship.

Conflict of interest

The authors declare no conflict of interest.

Author details

Girish Chandra*, Ujala Rani, Birkishore Mahto and Gopal Kumar Mahato
Department of Chemistry, School of Physical and Chemical Sciences, Central University of South Bihar, Bihar, India

*Address all correspondence to: girish@cub.ac.in

References

[1] Jerca FA, Jerca VV, Hogeboom R. Advances and opportunities in the exciting world of azobenzenes. Nature Reviews Chemistry. 2021;**6**:51-69. DOI: 10.1038/s41570-021-00334-w

[2] Patel S, Patra S, Chand S, Singh RK, Singh I, Chandra G. Chemical and photophysical studies on the interaction of reactive oxygen species with carbonylated diazo derivatives: Their chromogenic, fluorescent detection and mechanistic investigation. Dyes and Pigments. 2023;**216**:111317. DOI: 10.1016/j.dyepig.2023.111317

[3] Patel S, Mahato GK, Chandra G. Unusual oxidation of methylated diazo dye to 4,4′-dihydroxy-azobenzene: Mechanistic investigations and spectroscopic studies of the isolated product. Dyes and Pigments. 2023;**218**:111497. DOI: 10.1016/j.dyepig.2023.111497

[4] Patel S, Das B, Mishra P, Chandra G. Light triggered reversible aggregation/dispersion of hydroxy azo-benzenes during photo switching: Solvent, ions assisted dispersion, and induced quenching emission. ChemPhotoChem. 2024:e202300350. DOI: 10.1002/cptc.202300350

[5] Modi J, Patel S, Mahato GK, Mahto B, Chandra G. Effective light-induced fluorescent H-aggregation of fluorinated pyridyl-azo dye via photoisomerization. ChemPhotoChem. 2024:e202300284. DOI: 10.1002/cptc.202300284

[6] Goulet-Hanssens A, Eisenreich F, Hecht S. Enlightening materials with photoswitches. Advanced Materials. 2020;**32**:1905966. DOI: 10.1002/adma.201905966

[7] Fedele C, Ruoko TP, Kuntze K, Virkki M, Priimagi A. New tricks and emerging applications from contemporary azobenzene research. Photochemical & Photobiological Sciences. 2022;**21**:1719-1734. DOI: 10.1007/s43630-022-00262-8

[8] Tecilla P, Bonifazi D. Configurational selection in azobenzene-based supramolecular systems through dual-stimuli processes. ChemistryOpen. 2020;**9**:538-553. DOI: 10.1002/open.202000045

[9] Oscurato SL, Salvatore M, Maddalena P, Ambrosio A. From nanoscopic to macroscopic photo-driven motion in azobenzene-containing materials. Nano. 2018;**7**:1387-1422. DOI: 10.1515/nanoph-2018-0040

[10] Broichhagen J, Frank JA, Trauner D. A roadmap to success in photopharmacology. Accounts of Chemical Research. 2015;**48**:1947-1960. DOI: 10.1021/acs.accounts.5b00129

[11] Wie JJ, Shankar MR, White TJ. Photomotility of polymers. Nature Communications. 2016;**7**:13260. DOI: 10.1038/ncomms13260

[12] Busschaert N, Caltagirone C, Van Rossom W, Gale PA. Applications of supramolecular anion recognition. Chemical Reviews. 2015;**115**:8038-8155. DOI: 10.1021/acs.chemrev.5b00099

[13] Chi X, Cen W, Queenan JA, Long L, Lynch VM, Khashab NM, et al. Azobenzene-bridged expanded "Texas-sized" box: A dual-responsive receptor for aryl dianion encapsulation. Journal of the American Chemical Society. 2019;**141**:6468-6472. DOI: 10.1021/jacs.9b01241

[14] Mahto B, Chand S, Singh DV, Patra S, Chandra G. Synthesis of hybrid

azobenzene-chalcone dyes and studies on their photophysical properties: Molecular recognition of carbonate and phosphate ions using a novel turn-off fluorescent probe. Journal of Molecular Structure. 2024;**1306**:137839. DOI: 10.1016/j.molstruc.2024.137839

[15] Renner C, Moroder L. Azobenzene as conformational switch in model peptides. Chembiochem. 2006;7:868-878. DOI: 10.1002/cbic.200500531

[16] Klajn R. Spiropyran-based dynamic materials. Chemical Society Reviews. 2014;**43**:148-184. DOI: 10.1039/C3CS60181A

[17] Berberich M, Würthner F. Terrylene bisimide-diarylethene photochromic switch. Chemical Science. 2012;**3**:2771. DOI: 10.1039/c2sc20554h

[18] Wang Y, Bie F, Jiang H. Controlling binding affinities for anions by a photoswitchable foldamer. Organic Letters. 2010;**12**:3630-3633. DOI: 10.1021/ol1014043

[19] Hossain MS, Rahaman SA, Hatai J, Saha M, Bandyopadhyay S. Switching the recognition ability of a photoswitchable receptor towards phosphorylated anions. Chemical Communications. 2020;**56**:4172-4175. DOI: 10.1039/d0cc00926a

[20] Iamsaard S, Anger E, Aßhoff SJ, Depauw A, Fletcher SP, Katsonis N. Fluorinated azobenzenes for shape-persistent liquid crystal polymer networks. Angewandte Chemie International Edition. 2016;**55**:9908-9912. DOI: 10.1002/anie.201603579

[21] Hansen MJ, Lerch MM, Szymanski W, Feringa BL. Direct and versatile synthesis of red-shifted azobenzenes. Angewandte Chemie International Edition. 2016;**55**:13514-13518. DOI: 10.1002/anie.201607529

[22] Hossain MS, Bandyopadhyay S. Metal ion mediated instant Z → e isomerization of azobenzene macrocycles in the absence of light. The Journal of Organic Chemistry. 2021;**86**:6314-6321. DOI: 10.1021/acs.joc.1c00105

[23] Ahmad M, Metya S, Das A, Talukdar P. A sandwich azobenzene–diamide dimer for photoregulated chloride transport. Chemistry: A European Journal. 2020;**26**:8703-8708. DOI: 10.1002/chem.202000400

[24] Hatai J, Bandyopadhyay S. Photoreversible assembly-disassembly of a polymeric structure by using an azobenzene photoswitch and Al^{3+} ions. Chemistry: A European Journal. 2014;**20**:10020-10026. DOI: 10.1002/chem.201402321

[25] Accardo JV, McClure ER, Mosquera MA, Kalow JA. Using visible light to tune boronic acid-ester equilibria. Journal of the American Chemical Society. 2020;**142**:19969-19979. DOI: 10.1021/jacs.0c08551

[26] Kerckhoffs A, Moss I, Langton MJ. Photo-switchable anion binding and catalysis with a visible light responsive halogen bonding receptor. Chemical Communications. 2022;**59**:51-54. DOI: 10.1039/d2cc05199k

[27] Dąbrowa K, Jurczak J. Tetra-(meta-butylcarbamoyl)azobenzene: A rationally designed Photoswitch with binding affinity for oxoanions in a long-lived Z-state. Organic Letters. 2017;**19**:1378-1381. DOI: 10.1021/acs.orglett.7b00267

[28] Sokołowska P, Dąbrowa K, Jarosz S. Visible-light responsive sucrose-containing macrocyclic host for cations.

Organic Letters. 2021;**23**:2687-2692. DOI: 10.1021/acs.orglett.1c00590

[29] Hua Y, Flood AH. Flipping the switch on chloride concentrations with a light-active foldamer. Journal of the American Chemical Society. 2010;**132**:12838-12840. DOI: 10.1021/ja105793c

[30] Wang Z, Cook MJ, Nygard AM, Russell DA. Metal-ion chelation and sensing using a self-assembled molecular photoswitch. Langmuir. 2003;**19**:3779-3784. DOI: 10.1021/la026792n

[31] Dabrowa K, Niedbala P, Jurczak J. Anion-tunable control of thermal Z→E isomerisation in basic azobenzene receptors. Chemical Communications. 2014;**50**:15748-15751. DOI: 10.1039/c4cc07798a

[32] Shinkai S, Ishihara M, Ueda K, Manabe O. On-off-switched crown ether-metal ion complexation by photoinduced intramolecular ammonium group tail-biting. Journal of the Chemical Society, Chemical Communications. 1984:727-729. DOI: 10.1039/C39840000727

Chapter 6

Distinctive Properties of Lasing with Dye-Doped Liquid Crystals at the Planar Alignment of a Helical Structure

Eugene A. Tikhonov and Igor P. Ilchishin

Abstract

This review focuses on the deliberate study and develop of the most atypical members of the laser family, which use dye-doped matrixes on nematic and cholesteric liquid crystals (LCs). The cause lies in their strengthened light scattering, which can be partially overcome by forced alignment and subsequent light amplification. Their attraction remains due to LC birefringence, which changes under a small outside electric field. Therefore, this review considers the lasing features of dye-doped LC lasers with tunable wavelength oscillations at recording flashy reflective Bragg gain gratings of variable spatial periods in crossed pumping beams. The second approach of lasing oscillation affords steady-state chiral reflective Bragg gratings. Lasing in both approaches is passably described in the frame of coupled wave theory. According to this theory reproducible one-mode oscillation occurs when modifying chiral gratings. Lasing on an induced nematic LC with deeper modulation index refraction revealed the additional spectral state of the LC: the state of a photonic crystal. The existing lasing models need to consider the attributes of the LC photonic crystal, such as the bandgap.

Keywords: distributed cavity, ordering of the scattering domain, dye-doped liquid crystal, pulse holographic grating, induced chirality, lasing

1. Introduction

The acronym "laser," as the chronological follower to a "maser" acronym, was proposed by pioneers to refer to the process of "light amplification by stimulated emission of radiation." The scientific and virtual progress of lasers has led to a separation of the content of the original meaning of the acronym since the laser as a device also includes an optical resonator, amplifying media, and an energy pumping source specific for their excitation. The presence of a resonator turns a laser from a light amplifier into a light generator. In this respect, the dry and, therefore, rarely used abbreviation of Russian origin OQG (optical quantum generator) turned out to be more exact. Meanwhile, the acronym acquired new shades of content and derived terms such as laser, lasing, and even the verb to lase. Readers will meet these terms on the following pages.

Liquid crystal lasers followed the bright time lap of dye lasers in conventional transparent liquids and polymeric solutions. The lasing thresholds of these dye lasers were surprisingly low (despite the great spectral width in comparison with that of ruby), and the prospect of coherent lasing radiation in the 350–1000 nm range was attractive. This range follows from the addition of luminescence spectral widths of the relevant dyes after the first pioneering publications [1–4]. For spectral tuning, even one 60-degree prism inside the Fabry-Perot laser cavity sufficiently overlapped the mentioned range. (I. Newton was the first who applied a similar prism to separate the primary colors from white light and back to white light.)

LC in a polycrystal state falls into a collection of low-symmetry domains with randomly oriented optical axes. Due to such spatial and optical inhomogeneity, the light scattering of similar layers compared to that of ordinary liquids is much greater, much greater than that of the molecular Rayleigh analog. Therefore, trying it in a dye-doped set of random lasers would be okay. The ordering of LC domains through local interactions with another ordered surface reduces this light scattering to a depth of approximately 60 μm. The technology of rubbing the inner sides of substrates allows to alignment of the separated domains in a uniform direction and results in a homogeneous transparent layer (50–60)μm thick. Meanwhile, the next heavy obstacle is proven afterward. At least for lasers with longitudinal pumping, it is necessary to have an active medium with an optical density of at least d = 2.5–3.5 ($d = \varepsilon(\lambda)\mu L$), where $\varepsilon(\lambda)$ is the molar extinction, μ is the dye concentration, and L is the thickness of the layer. It is difficult to reach the indicated amount d at these values L when the critical dye concentration $\mu \geq 10^{-3}$ mol/l, at which point luminescent quenching (the quantum yield decreases) begins.

In the pioneer article published in 1976, lasing features on a dyed homogenous LC nematic in longitudinal pumping geometry showed quite simple results in the shadow of impressive advances already demonstrated by liquid dye lasers [5]—the reason for which resided in limitations such as a small active lasing volume for longitudinal laser pumping. Instead of the limited transparency in depth, even for an aligned LC of approximately 100 microns, a better solution can be achieved with the geometry of transverse pumping. The problem of a new boundary conditions affected by the wavefront of oscillating beams appears. The outside optics like semiconductor lasers and can be used to solve this problem.

The following parts of the review are intended to focus on wavelength-tuning methods conjugated with a spectral width narrowing of lasing emission conformable to that of usual dye DFB lasers and two types of LCs-dyed nematic and helical structured lasers.

2. Liquid crystal lasers with dynamic distributed feedback

Typically, lasers are based on optical cavities (resonators), while stimulated emission at high amplification can develop in running wave mode without a cavity, which is called superluminescence, with slight spectral and angular narrowing. The resonator is responsible for lasing on its own modes under known conditions in the following form: gain ≥ losses or $\alpha(\lambda) \geq g(\lambda)$. Positive feedback occurs when every loop transitions the emission through the amplifier and cavity, resulting in an increase in the intensity of the cavity modes and a decrease in the luminescence power. When the initial gain satisfies the deep inequality $\alpha(\lambda) \gg g(\lambda)$, luminescent noise can often favor a quick rise to superluminescence out of loop transits, which are necessary for cavity modes [6].

The laser cavity is the well-known interferometer Fabry-Pero that was long already used in high spectral resolution spectroscopy. When the distance between its two mirrors equals L and the refractive index is "n," the optical length is Ln, so the value $2Ln/\lambda = m$ is an order of magnitude of interference. The greater the value has "m," the greater the spectral resolution of an interferometrical measurement. Moreover, in diffraction measurements, the spectral resolution is determined by the total number of equivalent elements involved in diffraction. Therefore, in this case, the number "m" determines the spatial modulation diffraction order, which may be considered as distributed feedback of the laser cavity (DFB). The theoretical analysis of such periodical structures in the context of thick phase holographical gratings H. Kogelnik was presented in 1969. A version of this theory for a DFB laser based on a reflective Bragg grating (H. Kogelnik) was presented in 1972 [7].

Let us proceed to a description of optical configurations that combine optical pumping and recording of the flashy volume gratings in the dyed planar alignment nematic LC. The first experimental study of dynamic DFB lasing on a homogenic aligned dyed nematic LC with transverse pumping geometry, which combined pumping functions and the flashy grating, was published in [8].

The optical schemes in **Figure 1C** and **D** provide examples of optical schemes operated in those regimes. Pumping/grating laser beams in both cases support together the needed power level and the rate of coherences that are necessary to build the periodic structure with a spatial period $\Lambda = m\lambda/2n$. Diffraction in the first order at $m = 1$ provides a greater efficiency than that in the subsequent orders at $m = 2.3$,

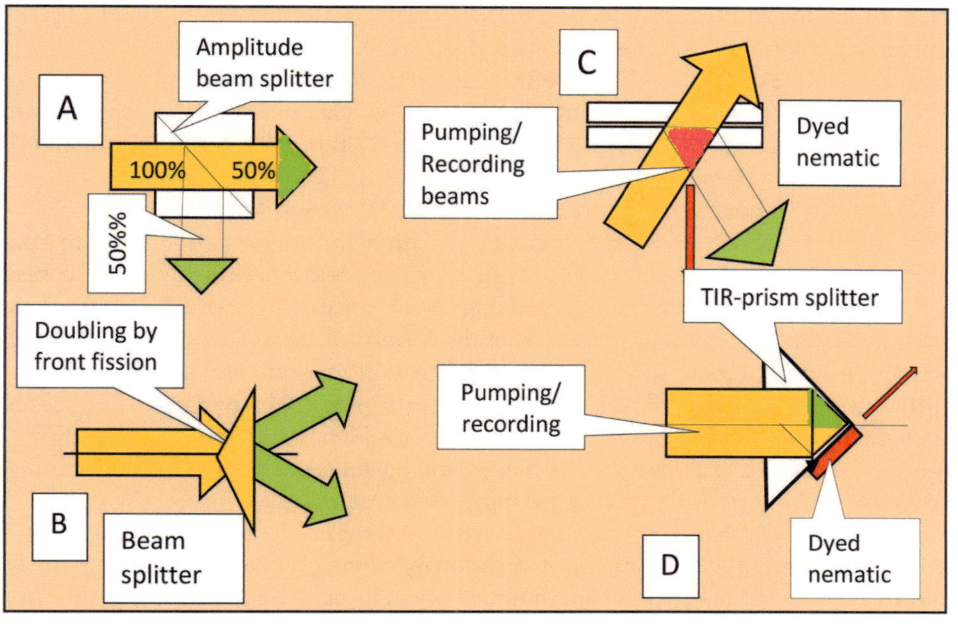

Figure 1.
(A, B, C, D) 1A—amplitude beam doubling by fission, 1B—beam doubling by separation on half-in-half of the wavefront—schemes of lasing with pumping/pulse grating together: 1C—longitudinal pumping and flashy grating record at the overlap of direct and reflected beams, and 1D—transversal pumping with TIR-prism.

at least at the harmonic profile of spatial modulation. Previously, it was shown that a one-mode pumping laser enables one-mode oscillation of a conventional dye laser using dynamic DFB [9].

Now some comments on the methods of beam doubling and interference schemes for the best excitation and building a flashy grating in laser media. This grating arises from locally excited dye molecules that create a spatially modulated gain. Therefore, lasing in this way can be achieved even with picosecond pumping [10]. When the pumping time gets longer of 10 or up to a 100 nanoseconds, local heating, and expansion due to dissipative transitions destroy the optical homogeneity and grating that disrupts lasing. Now, it is clear why it is critical to realize a bright interferogram and to record an efficient grating for lasing with tuning. No problem arises when running on a one-mode pumping laser. However, virtually, solid-state lasers often operate with low spatial coherence on the average. After doubling even with amplitude fission mirror, beam copies cannot build up an interference pattern of the highest contrast without some conditions. It is clear without further ado for the interference of beam copies of the same pumping beam fission across the wavefront. Only overlapping amplitude copies in the Fresnel diffraction zone will provide the highest interferogram contrast when optical schemes permit the crossing of two direct images of both beams. **Figure 1** presents the various doubling pumping/recording schemes. **Figure 1A** and **B** show the doubling of the pumping beam without recording the dynamical grating. However, flashy virtual gratings may perform a beneficial wavelength-tuning function in the dye-doped laser family. Our experiments showed that tunable lasing free from an "escort" of waste superluminescence was inevitable.

Figure 1A provides much better results by improving the contrast of interferential pattern from the pumping/recording beam at overlapping beam copies of direct and inverted fronts. **Figure 1A** shows that the amplitude doubling clarifies how to reach a greater contrast of interference between two beams with fore wavefronts: it is sufficient to set one additional reflection in that scheme. The total repression of superluminescence in this way has been achieved [11].

As shown in **Figure 1C**, a 100% reflecting mirror doubles the pumping beam through the fission of the wavefront. The red-colored part points to the region where were overlapping direct and attenuated (it is a defect) reflecting beams occur, and the red arrows show a trend of lasing emission.

Now, it is clear how critical it is to realize a bright interferogram and record an efficient grating, which are tightly connected. No problem arises when running on a one-mode pumping laser. However, there are often multimode lasers with low spatial coherence. Then, the good copies through doubling such beams are performed with amplitude fission and beam separation on two parts of the wavefront. This is because the following overlapping amplitude copies in the Fresnel diffraction zone will provide a higher contrast interferogram than local copies when overlapping coherent spots are impossible in general. Two doubling pumping/recording beam schemes are shown graphically in **Figure 1A** and **B**. They show optical schemes suitable for more homogeneous pumping dye lasers (without recording flashy gratings). However, flashy virtual gratings may perform a beneficial wavelength tuning function in the considered dye laser class. Our experiments showed that tunable laser emission by varying the interference angle free from an "escort" of the needless superluminescence emission was attainable in this case.

Figure 1C shows that doubling the pumping beam provides a 100% reflective mirror. The red-colored part of the scheme points to the region where cross-fore attenuated and back reflective beams occur. The lasing emission λ_{osc} follows from the known relation for phase conditions:

$$\lambda_{osc} = \lambda_p n / \left(n^2 - \sin^2\theta\right)^{1/2} \tag{1}$$

The lasing wavelength λ_{osc} of a DB laser in the scheme in **Figure 1C** is a function of the phase oscillation condition: $\Lambda = \lambda_{osc}/2n$, where λ_p is the wavelength of the monochromatic pumping beam and θ is the incident angle of the nematic LC layer such that the variation in the refractive index n of the LC affects the lasing wavelength. An electrical switching LC director from a homogenic to homeotropic orientation allows wavelength tuning. As we emphasized earlier, the severe lack of longitudinal pumping resides in the limited gain length due to the limited depth of the ordered LC layers. Because the lasing threshold is a function of the amplification "αL," an optimal density of 3–4 is not attainable. The second lack of the short length "L" resides in the restricted gain and laser energy.

Application to transverse pumping geometry when the gain length "L" depends only on the cross-section of the pumping beam removes this drawback. This laser scheme also solved the lasing problems of frequency-tunable picosecond pulses with picosecond pumping due to the synchronous time crossing of paired beam copies recording a gain grating [10]. Moreover, considering the nature of a dynamic grating in picosecond time provides substantial evidence in favor of a flashy gain grating with an instantaneous response.

The lasing scheme with a TIR prism (**Figure 1D**) is famous for ordinary tunable dynamic DFB lasers in conventional solutions [12]. This scheme again offers a transverse pumping geometry that is sufficiently useful for LC lasing because of its unlimited gain length. The lasing wavelength complies with the conditions of first-order Bragg diffraction or phase oscillation:

$$\lambda_{osc} = n\lambda_p / n_{pr}\sin\theta = \sqrt{2}n\lambda_p / \left[\left(n^2_{pr} - \sin^2 i\right)^{0.5} \pm \sin i\right] \tag{2}$$

where n_{pr} is the refraction index of the TIR prism and "i" is the deflection angle. The main drawback of this scheme is the doubling of the pumping/recording beams by fission of the initial pumping beam across its wavefront.

Now, let us discuss such a magnificent laser feature with distributed feedback, which comes from mode selectivity regarding the higher axial modes. This critical feature repeatedly emphasizes the theoretical works of H. Kogelnik. However, experimental works devoted to LC lasing did not even address this problem until it was recognized actual.

According to [7], the threshold amplification of a DFB laser at lasing frequency ω is determined by the next expression.

$$\exp(2\alpha L) = 4\alpha^2 / M^2 + m\Delta\omega^2 / c \tag{3}$$

with coupling degree $M = \pi n_1/\lambda + j\alpha_1/2$ and where n_1 and α_1 are amplitudes of the modulation of the refractive index and gain values and the "Δω" frequency interval between axial modes or detuning from the exact Bragg frequency. It indicates clearly that modes m > 1, 2, 3... are inherent in the growing excitation threshold. Overall, relation Eq. (3) describes the more complicated lasing with the inclusion of a few parameters compared to the critical condition of self-excitation for lasers with conventional mirror cavities. For example, pumping in nanosecond time intervals creates conjoint grating parameters such as modulation of the gain α_1 and a basic gain α. The relation between them virtually depends on the contrast of the interference pattern that appears when crossing two copies of the "mother" pumping beam. The nature of the grating depends on the relationships among the refractive index modulation n_1, gain index α_1, and complex modulation $(n_1 + j\alpha_1)$, which define various lasing characteristics. In the approximation of one-mode oscillation, m = 0 when (1) $\alpha \gg M$ or (2) $\alpha \ll M$ in Eq. (3) is used to determine specific situations. In case (1), Eq. (3) is as follows:

$$4\alpha^2 \exp(-2\alpha L) = \alpha_1^2 / 4 + (\pi n_1 / \lambda_0)^2 \tag{4}$$

In this approximation, oscillation is sufficient for creating any gratings. If the deviation oscillation frequency from the wavelength interval equals neighboring axial modes, the coupling value M must double to reach self-excitation conditions. This means that the oscillation threshold in this mode will increase with the value of "$\alpha\lambda^2/2\pi n$." In the approximation $\alpha \ll M$, two cases exist: (a) M is the real number due to the spatial modulation of the refractive index, and (b) M is the imaginary number at the spatial modulation of the gain caused by pumping. In the case (a) mode m = o is impossible, the modes for m = ±1 shifted around λ_0 up and down $\pi n_1/\lambda_0$ and have an oscillation threshold of $\alpha L \approx (\lambda/n_1 L)^2$. However, in case (b), oscillation on of the resonance Bragg wavelength λ_0 is allowed at the threshold condition $\alpha_1 \approx \pi/L$.

These simplified theoretical evaluations of the thresholds and mode selectivity of DFB lasing are critical for the obtaining a qualitative understanding of the observed experimental results for dyed LCs and conventional liquid/solid lasers. For example, a liquid flashy DFB dye laser under a pumping by one-mode nanosecond pulses oscillates one-mode pulse emission. However, if the pumping/recording emission is changed to multimode one, then DFB dye laser emission also becomes multimode [9]. It looks like multimode emission records a few different gain gratings with other spatial periods.

Similar behavior allows the gain grating to be considered the critical feature of flashy DFB lasing despite the low dye concentration compared to that of the solvent. The phase thermal grating may arise and increase through the heating of the solvent (like liquid, polymer, or LC) in time of pumping: from low $(\partial n/\partial T)_v \Delta T$ to slowly growing $(\partial n/\partial T)_p \Delta T$. The first member does not carry in the sensible deposit to refraction index of the laser matrix, but the second member is strongly affected by the sound velocity in the given media. Therefore, a pumping pulse of (10–30) ns carrying less than one Joule has no time to create a predominant phase grating over the gain.

Direct measurements the time of appearance and relaxation of the thermal lenses were conducted in our previous publication [13]. This lens was the direct analog of a pumped region in dyed liquid and polymer DFB lasers. These measurements did not reveal any thermal lenses during the heating time of pulse pumping (20–30) ns but revealed their appearance and continued to track their expansion up to a steady state after τ = (1.3–2) μs. These measurements showed that polymeric matrixes are near on magnitude but more resistant than liquids to similar shock thermal perturbations under the same development time.

3. Liquid crystal lasers with steady-state natural chiral gratings

Research on starting LC lasing based on chiral LC structures [14] was published in 1978. Our motivation for this work was to investigate natural periodic structures with deep-phase modulation for linear and nonlinear optics and optical waveguides.

The chiral LC has a temperature-dependent patch of chiral structure "P" in the broad spectral range, including the visual interval. Like a nematic LC, a chiral LC structure is a self-ordered domain with randomly oriented of optical axes called a confocal system. It is possible to upgrade the confocal state to a planar state at a tangent shift of orienting substrates at a thickness less than about (100–200)μm. Light scattering after alignment is reduced in some order but remains stronger than that in trivial liquids.

Two groups of chiral LCs are broadly known today. Cholesteric LC (CLC) originated from a mixture of cholesteryl esters. The system transforms into a nematic LC if the right-handed and left-handed ester concentrations are equal. A CLC with a right- or left-twisted spiral arises if concentrations are unequal. The second group included the induced nematic group, which received CLC under an admixture of optically active ingredients.

The linearly polarized light transmitted through the aligned planar layer of such a CLC grating becomes circularly polarized due to the CLC birefringence. If the wavelength of circularly polarized light satisfies λ = P and P/2 = Λ, the polarization twists of light and the twist of the CLC are the same, and the light meets the Bragg conditions and is efficiently reflected. In other words, this CLC once satisfied the steady-state DFB lasing requirements. Optical parameters like CLC include an average refractive index $n = (n_o + n_e)/2 = 1.5$ and a value $\Delta n = (n_e - n_o)$, which determine the spectral parameters of the Bragg reflection band and, and particularly depth of modulation.

In the first group of CLCs, the modulation depth is $\Delta n \approx 0.045$. The second group can change Δn widely ~(0.1 ÷ 0.2). This manifests in the distinctive properties of lasing and its spectral characteristics in these two groups of LCs.

An estimation of the bandwidth reflection of the similar phase grating between the first two zero points of the Bragg reflection profile provides the following relation [15]:

$$\Delta\lambda = \lambda\Delta n \quad (5)$$

For a wavelength of λ = 560 nm, Eq. (5) provides a value of ≈25 nm, which is close to the experimental profile presented in **Figure 2**.

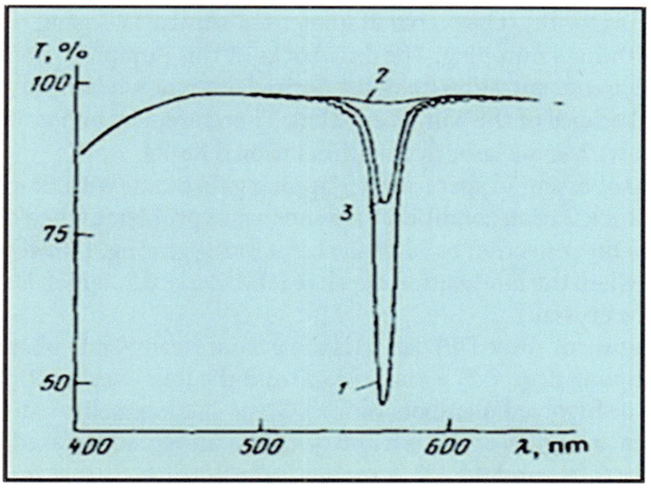

Figure 2.
Transmission spectra of the chiral grating at a thickness of 15 μk, T = 22°C, normal light incidence of 1—left circularly polarized light, 2—right circularly polarized light, and 3—linearly polarized light [14].

The studied laser configuration contains a chiral grating as the reflection mirror, whose reflection profile is presented in **Figure 2**; a standard dielectric 30% reflection mirror and an R6G dye solution cell under the Brewster angle. The pumping geometry satisfies longitudinal excitation by the second harmonic 532 nm of the Nd^{3+}-laser. Operating with circularly polarized laser emission is necessary to reach maximum reflection from the chiral CLC grating. For this purpose, in front of a left-twisted reflection grating, a λ/4-plate is used. A wavelength tuning of 567–585 nm is attainable in the taken configuration under a temperature change in the range of (22–28)°C.

We briefly discuss the other laser schemes that are feasible for circular lasing modes with phase chiral gratings as mirror.

1. Laser with two equal CLC mirrors with the same twist and isotropic gain of the dye liquid cell. Oscillation is excited on a circularly polarized wave that both mirrors support.

2. Its gain is anisotropic and supports linearly polarized emission. If both CLC mirrors have a left-handed twisted spiral, the oscillation excites on the left-handed circular polarized, and the right-handed circular polarized emission runs out of the cavity.

3. Instead of the one chiral grating, we put a conventional dielectric mirror, a sign of circularly polarized light changes at a reflection on the opposite, which transmits through the left-twisted mirror. The plates were oscillated until the λ/4 plate was reached, and the plates were installed as described above.

Considering this scheme, one can observe the finite ability (just temperature) of frequency tuning to support similar CLC mirrors. In addition to temperature frequency tuning and impossible tuning by angular scanning of any new features, this grating does not consist of nonzero angles of reflection. Therefore, let us pay attention to the more convenient lasing technology of steady-state distributed feedback with spatial modulation of complex index wave propagation that is popular today: cholesteric LC and induced nematic LC.

This approach's experimental study of lasing originated from the following pioneering works [16–18]. The planar alignment of similar LC lasing structures leads us to longitudinal pumping. The drawbacks of this pumping geometry include increasing the dye concentration up to the desired amount when the dye quantum yield decreases because of the pair interaction. Therefore, it is impossible to reach the lasing threshold with some laser dyes within rational limits.

The presence of essential spectral overlapping gain bands with Bragg reflection bands under self-excitation conditions does not cause problems when dye gain bands are broader than the reflection band of the helix Bragg grating. (However, this may become untrue when the modulation depth reaches $\Delta n \geq 0.1$, which is followed by the rise of a photonic crystal.)

The main feature of these DFB lasers is the partial steady-state phase character of the complex coupling degree $M = \pi n_1/\lambda + j\alpha_1/2$ and the following relation between the magnitudes of the involved members $(\pi n_1/\lambda)^2 > (\alpha_1/2)^2$. Inequality can be confirmed numerically because values of $n_1 = \Delta n$ and α_1 can be measured for used cholesteric and nematic LCs: 0.045 and 0.1 ÷ 0.3, respectively. We investigated the spectroscopic properties of lasing media in the case of cholesteric LC dissolved in 3-dimethylamino benzanthrone dye. Its absorption band overlaps wavelength pumping at λ = 532 nm. The planar-oriented CLC layers have a thickness of L = 40 μm and a dye concentration of 10^{-2} mol/l. The deformed spectral absorption, fluorescence, and reflection spectra of the CLC grating are presented in **Figure 3a** and **b**.

A deep in fluorescence band may be the first direct observation the photon confinement. All the observed changes in the spectra of the separated components of the combined media manifested because of their interactions. A given composition may be a typical representative of cholesteryl derivatives: cholesteryl oleate + cholesteryl palagonite + cholesteryl chloride. The presence of cholesteryl oleate ensured that the mesophase was present over a wide temperature range, from 10°C to 70°C. The initial selective-reflective wavelength and the rate of temperature tuning of the helix pitch caused the percentage of cholesteryl chloride in the mixture. To realize CLC with a mesophase temperature interval from 10°C to 65°C and a selective-reflective tuning speed of 2–4 nm/deg., the following mixture was prepared for application: 40% cholesteryl oleate, 35% cholesteryl palagonite, and 25% cholesteryl chloride. The alignment of the CLC texture in one direction produces the tangential shift of two rubbed before the substrates as much as possible. To stabilize the whole texture at layer thicknesses above 80 μm, aging at multiple melting and slow cooling to the LC state should be done.

The choice of a laser dye for activating CLC is difficult because of its small dissolving ability. Low or limited solubility and dimerization decreases the quantum yield and counteracts the development of dye-doped LCs for this purpose.

Like usual laser dyes but more relevant to dyes for LC lasing, they should have a great Stokes shift between the absorption and fluorescence bands. In this case, the profiles of the gain and fluorescence dye bands become similar, and a laser might have a low oscillation threshold. The dyes from the chemical family of 3-aminobenzanthrone derivatives conform to the primary claims of LC lasing.

Now, let us return to new spectral features of dye-doped LC media prepared for lasing on circularly polarized waves under conditions of DFB CLC. It is critical to determine the nature of the deep signals emerging in the profiles of the fluorescence and transmission of a dye around the reflection band of Bragg grating CLC (**Figure 3a** and **b**). The profile (dashed line) of the reflection band in the

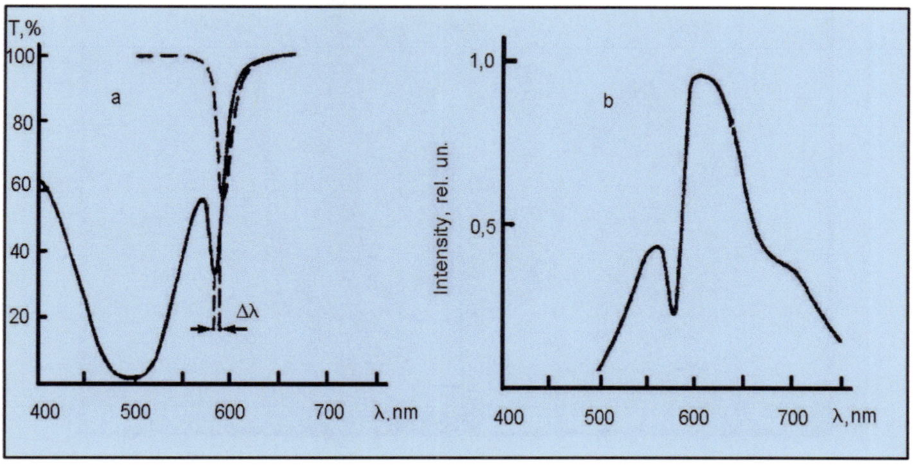

Figure 3.
(a) Transmission (T) spectra of pure cholesteric liquid crystals (dashed line) and CLCs activated with 3-dimethylamino benzanthrone dye in linearly polarized light and (b) fluorescence spectrum of dyed CLCs of the same thickness for left circularly polarized light [17].

plane-aligned CLC without laser dye is shown in that figure. Deeps in its spectra arise when the dye dissolves in CLC due to the interaction of spiral grating CLC with absorption and fluorescence transitions. The depths in both profiles reveal a spectral shift of ≈4 nm in their maxima. The discovery of these features in terms of absorption and fluorescence is possible only with left circularly polarized light. It is well-known that both components, left-handed and right-handed circular light, make linearly polarized light.

However, more importantly, this finding strongly proves that helicity-aligned CLC interacts with solute dye molecules and builds up them likewise in helicity alignment. Therefore, there is some increased absorption for the Bragg wavelength and the narrow spectrum around it. These dye molecules have conserved their helix-twisted alignment in an excited state. This part of the circularly polarized fluorescence is captured by the distributed CLC resonator. The fluorescence measurement results must be considered because of the mutual orientation of the bodily angle relative to the axis of the chiral grating. Only the parallel orientation of the axes in these two directions allows for the recording of deep spectra, as shown in **Figure 3b**. However, why do we see a deep spike? All other parts of these fluorescence spectra suffer a decrease in the distributed power, even though the amount of power required for registration by our device is too small.

Conversely, a deep inside of the fluorescence spectrum appears in the reflection band at the wavelength of lasing emission near the wavelength of the maximum fluorescence band (**Figure 4**). This figure shows typical conditions before a lasing-conditioned DFB CLC. This process repeats the preceding result but shifts the absorption wavelength peak of the dye to the pumping wavelength. This state develops as an exit before being observed deep in the absorption profile and at an optimal position close to the maximum fluorescence band. The reabsorption effect is responsible for the Bragg reflection band's registered ≈4 nm shortwave shift (**Figure 3a**).

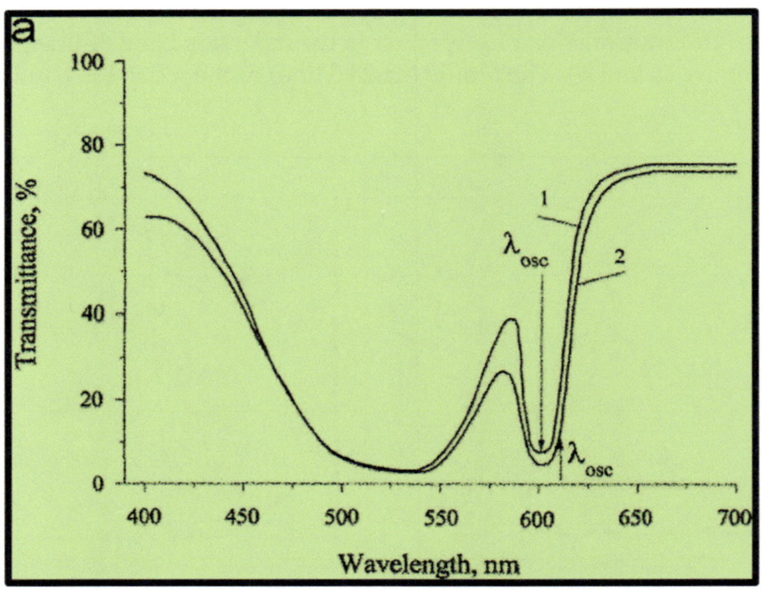

Figure 4.
Transmission spectra of dye F490 in a mixture of CLC with the following components: 40% cholesterol palagonite, 25% cholesterol chloride, and a 1—glass substrate with SnO_2 and polyimide coatings—glass substrate with a polyimide lacquer layer; measurements were performed under circularly polarized light with a thickness of L = 45 μm. The arrows indicate the positions of the lasing spectra [17].

The next part of this work will examine the main characteristics of lasing emission after achieving the oscillation threshold [18, 19]. First, for the best understanding of specific DFB system lasing, let us compare the modes of a helix phase reflective grating with L ≈ 40 μm and a base based on a classical cavity type Fabry-Pero with a length of L = 25 sm.

The distance between two neighbors axial modes can be calculated for both devices from the relationship $\Delta\lambda \approx \lambda^2/2\,L$, which provides a value of 7.2×10^{-6} nm for the classical cavity, and $\Delta\lambda \approx \lambda_0^2/2\,L = 4.5$ nm for the twisted reflective grating. The reflection bandwidth of dielectric mirrors is much broader than that of gain bands, even for dyes; therefore, no restrictions exist for many oscillating modes inside a gain bandwidth, including saturation gain.

The observed number of axial modes of the DFB CLC laser in repeated experiments was 3 (**Figure 5a**). Arrows indicate the positions of lasing spectra in a system of (P/P_{thr}, λ). The image presents sequential flash photos of lasing spectra with subnanometric resolution as a function of the relative pumping power P/P_{thr}. They have registered mode intervals from **Figure 5a** $\Delta\lambda \approx 5$ nm, and the position of the central mode λ = 604 nm related to a peak Bragg reflection allows us to check how their lengths L might agree. The actual length of the distributed reflection cavity equals L = 40 μm, while the experimental length from the known relation is $L_{dfb} = (\lambda_0)^2/2\Delta\lambda$ = m or ~72 μm.

In theory, Ref. [7] has shown that considering dispersion relations, the spectral position of lasing modes can differ from the results in the frame of a simple geometrical model.

Similarly, the unique method of realizing one-mode lasing with the considered type of DFB laser is presented in **Figure 5b**. In this case, one-mode operation has conserved a broad interval of pumping power [19, 20].

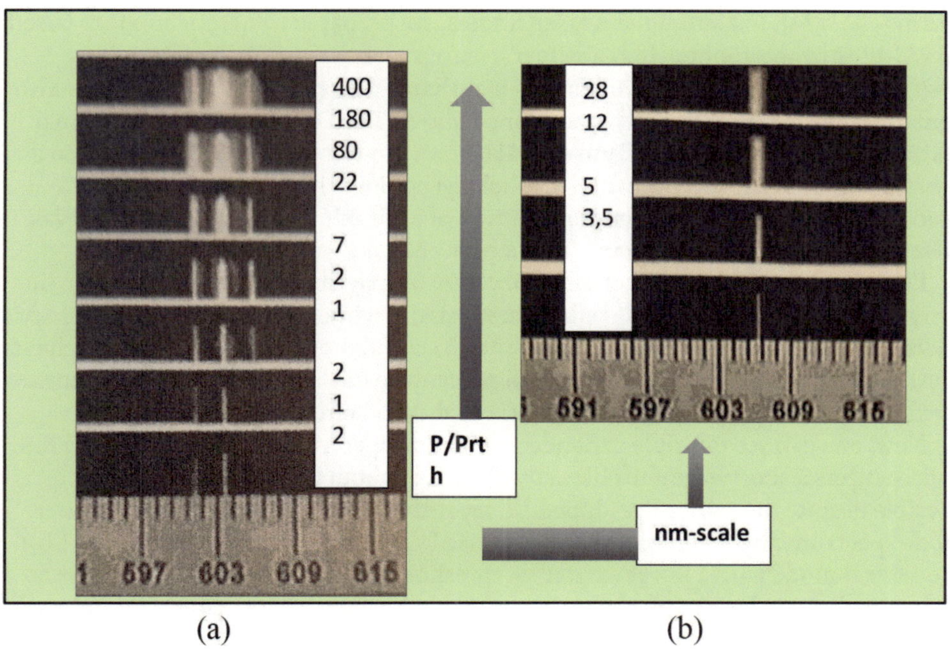

Figure 5.
(a) Alteration of oscillating mode spectra under pumping power at parallel director orientation, (b) one-mode oscillation and its conservation with pumping at mutually orthogonal director orientation [19, 20].

There have also been attempts to create defects in spiral gratings that respond such as "deep" inside restricted reflection spectra [19]. In this way, our twisted phase grating would respond to a definite requirement of the photonic crystal (oscillation on defective modes was already discovered). This approach was unsuccessful, but the tested method helped achieve one-mode oscillation with emission of the narrowed lasing line of the about tenth part of 1 nm. This method involves the application of crossed directions of rubbing on both substrates of LC cells.

Considering crossed-beam pumping lasing, we compared the spatial modulation of complex refractive indices with steady-state helical modulation for naturally twisted CLC devices under DFB conditions. The modulation depth Δn of the refractive indices in both laser media changed from ~0.001 to ~0.04 and was insufficient to manifest the evident features of the photonic crystal at lasing. However, the phenomenon of fluorescence confinement for dye-doped CLCs with natural helical structures is already a feature of this type (**Figures 3b** and **4**). Therefore, our next step in the past has been performed correctly because the transition to an induced nematic with a stronger modulation depth Δn might promote the development of photonic crystal attributes. The modulation depth for the induced LC follows the birefringence magnitude, and for an appropriate nematic may lie in the range of 0.1 ÷ 0.2. However, one should not forget that this parameter proportionally increases the spectral bandwidth of the reflection band and inhibits the realization of the optimal length of lasing media [7].

Lasing on the dye-doped induced LC with multiple broader reflection bandwidths due to more extensive modulation, Δn suddenly oscillates, which is different from the typical conditions of self-excitation (gain ≥ losses) for some wavelengths, as confirmed by our CLC laser [20]. Our initial explanation of this principal distinction based on the idea of two spiral gratings [18] was untrue despite specific features such as fluorescent confinement [21], leading us to the photonic crystal. Now, we look at the lasing behavior of dye-doped induced LCs with deep phase modulated gratings [22–24]. For subsequent laser studies, we prepared a eutectic mixture based on LC-1282 cyano-biphenyls as a nematic matrix and as the twisting ingredient (8–9) wt% 2-(4-phenyl benzylidene)-1-menthane-3-one additive. The mixture also contained the 0.5 wt% laser dye mentioned above. The DFB LC exhibited a planar texture at a thickness of L = 15 μm and a layer of dye-doped LC encased between glass substrate-coated rubbed polyvinyl alcohol layers. Both the spectrophotometer and spectrofluorometer had a spectral resolution of approximately 0.5 nm. Lasing was observed and studied under nanosecond pumping at a wavelength of 532 nm.

Figure 6a shows the transmission spectrum of a twisted LC with laser dye in linearly polarized light (1) and its reflective spectrum without dye at 8.12% optical active additive (OAA) for left- (2) and right-hand (3) circular polarization. The LC helix has a lefthand rotation direction, but a small percentage of the righthand component also attends and shows a weak reflection on the righthand circular polarization.

Now, one can see multiple expanded reflection spectrum widths of about 50 nm. It also suffers a shortwave shift like a dyed CLC of about 6 nm. DFB lasing was possible despite the minor dye-doped LC layer thickness of 15 μm. The oscillation mode spectrum also resembles the cholesteric LC of three axial modes (**Figure 5b**): the central mode with a lower excitation threshold and the two following modes at higher excitation. **Figure 6b** shows the transmission spectra for states with induced helicity in linearly polarized light (1, 2) and the gain spectrum of dye (3) (the gain and fluorescence profiles are similar). At OAA content of 8.12% (dashed curve 1), an oscillation line with a threshold of ~150 kW/sm^2 appears at λ = 582.4 nm on the shortwave edge of the expanded reflection band.

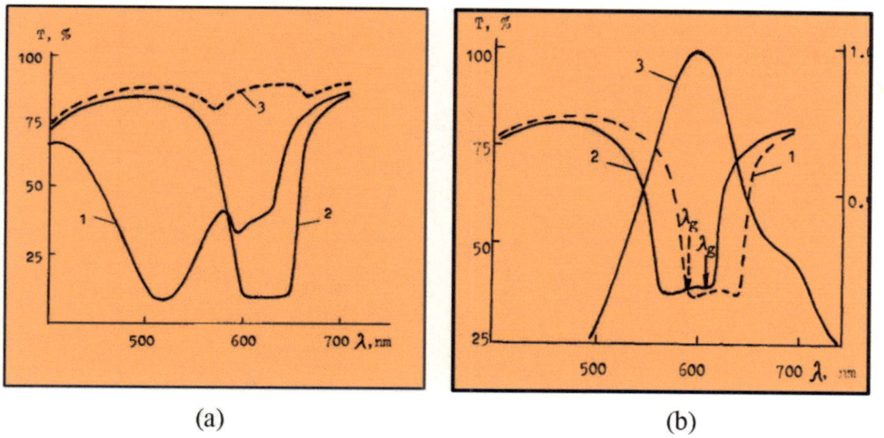

Figure 6.
(a) Transmission spectra of ZhK-1282 + 8.12% OAA with the dye (1) and without it for lefthand (2) and righthand (3) circularly polarized light, and (b) 1, 2—transmission spectra for ZhK-1282 at two various OAA contents (without dye), 3—fluorescence spectrum of impurity dye in LC [24].

Moreover, the very profound depth of the distorted reflective profile of the given system also points to this wavelength (**Figure 6a**, curve 1). With increasing OAA content up to 9%, the Bragg reflective band shifts to the short-wavelength edge of the spectrum (**Figure 6b**, curve 2). The lasing wavelength of the other grating jumps to 605 nm on the long-wavelength edge of the extended reflection band. The oscillation threshold, in this case, is ~50 kW/sm^2. Therefore, both lasing lines arose on opposite edges of the extended reflection band, as confirmed by many researchers. Therefore, similar oscillation lasing on "edge modes" is accepted. Their origin may also be related to Bragg gratings on opposite edges of the reflection band. The relation between the values of these wavelengths presented shortwave and longwave edge modes of our photonic crystal cavity equals 0.963.

Our work [18] demonstrated simultaneous oscillation on both edges of the reflective band realized with another nematic LC, LC-654, plus 36% cholesteryl oleate as a twisting component (OOA).

However, the newly expanded reflective band remained unaltered. The registered oscillation frequencies had 599.2 and 640 nm values, and their ratio also equaled 0.936. Therefore, these ratios differ from each other only in the second sign, which supports the following observations: the first is about oscillation on edge modes; the second is that the spatial period of grating changes the phase oscillation condition in the limit spectral width, thanks to the index refraction dispersion $\lambda = 2n(\lambda)\Lambda$; and the third is that oscillation inside two edges is unattainable. The new feature of chiral Bragg gratings arises with a deep index modulation of 0.1–0.2.

Here, the interaction between two oscillation models for the considered laser systems is time to consider. Model laser oscillation with inclusion features of photonic crystals has been proposed [24–27] to explain the ban of oscillation in the limit of the reflection band, considering it as the bandgap in photonic crystals. The oscillation threshold in this forbidden region is higher than that on the edges of the given gap. Oscillations on the edge modes are excited due to the opportune dispersion dependence of index refraction for given media and the axial modes described by Kogelnik's theory [7].

Although an arbitrary periodical structure (or Bragg grating) could resemble a photonic crystal, the bandgap, like its real feature, can exist only at enough profound

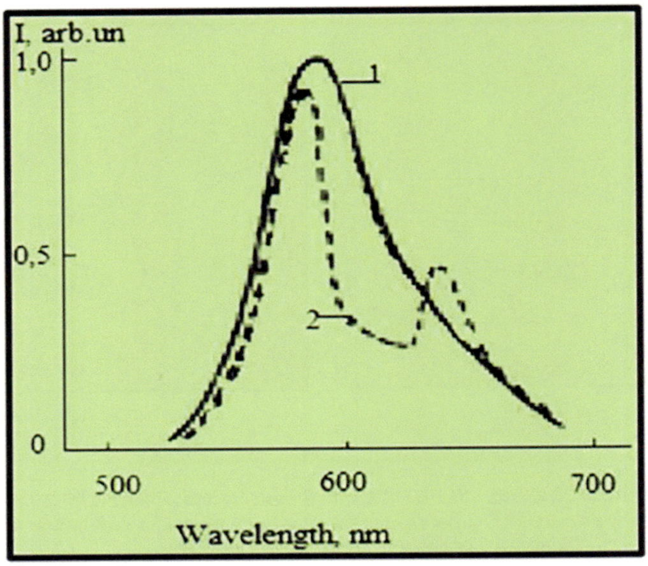

Figure 7.
Fluorescence spectra of dye-doped induced LC (the mixture LC-654 + 36% of Tecon-20): registration at an angle of 45° concerning the helix axis (1); and registration along the helix axis for the left circular polarization (2). The LC layer thickness is 30 μm [23].

modulation of index refraction. The bandgap was developed as forbidden for the excitation region of frequencies (energy) among other allowed regions. The discovery of deeps in a fluorescence band with an internal Bragg grating causes photon confinement and a definite spectral position of oscillation wavelengths. However, it is not yet a very photonic crystal. Light at these frequencies can be reflected or transmitted by compounds. Then, we focused on induced nematic media with a profound modulation depth $\Delta n \sim (0.1–0.2)$. The forecasted expansion of the reflective width unexpectedly resulted in new lasing behavior and the inability to tune the continuous oscillation frequency by changing the helix pitch.

In conclusion, let us think about confinement content is shown in **Figure 7**.

Continuous line (1) and dashed line profiles are trivial for spatially homogenous fluorescence emission and emission along helix axes (2). Both intensities are normalized to 100%.

A profound amount of deep intervention occurs on the right edge of the bandgap (2). It forecasts the relevant oscillation wavelength. A similar shortwave depth for shortwave spectral position reflection can be achieved with a variation patch of a helix or gain profile, as shown in **Figure 6a**. The side spikes show an increase in the intensity of the band gap.

4. Conclusions

The presented work has embraced long continuous intervals since the beginning of lasing investigations by applying various dye-doped LCs. Because fluorescent dyes for LC are not yet available for manufacturing, real-time investigations of dye-doped LC lasers have been performed on single casual samples or some dyes created early for lasers on usual liquid and polymer matrixes.

Despite the apparent deficiency of LC as a matrix for lasing on account of distinct types of light scattering (even in total alignment), infrequent attempts started in the 1970s. Division 2 has been shown using dye-doped aligned nematic crystals in optical schemes with laser pumping and simultaneous pumping/recording of flashy Bragg gratings. Moreover, this lasing method should always be combined with a primary advantage LC, namely, the simple ability to change the refractive index and the ability to readjust the direct LC with an outer electrical field. However, until recently, this advantage has remained unrealized.

The most successful approach for today (division 3) may consider the lasing of dye-doped LCs with natural and induced nematic chiral gratings. Due to the use of chiral Bragg gratings and gratings with variable spatial periods, distributed feedback cavities with spectral selectivity will have the ability to realize frequency-tunable oscillation. Thus, we can change the rate of monochromatic lasing emission by changing the modulation depth of the refractive index, which is responsible for new features of Bragg gratings. Moreover, at an enhancement of the modulation depth above of ~10% the refractive index, the transition from the usual dye-doped LC to a photonic LC with bands open and forbidden for light propagation or the transition to photonic crystals becomes possible.

The interaction of light with dye-doped photonic LCs with the bandgap provides new features that are unusual for prominent lasing media. Therefore, an LC with a bandgap permits oscillation only in separated spectral parts of limited reflection, such as edge modes on the edges of the bandgap, defect modes into the boundary of the bandgap, and others. There have been attempts to explain edge modes such as oscillation on some distortions of chiral gratings that make the creation of two gratings with different spatial periods [18, 22, 23].

However, the regular oscillations of edge modes under different conditions with various LCs cannot support this specific explanation. The oscillation begins on one of two bandgap edges or from both edges at different wavelengths but never inside the gap. The phase conditions of oscillation Eq. (2) for different wavelengths on two edges are satisfied only at the cost of the medium-index refraction dispersion. The mode spectra are nearly the same because they have a small modulation depth when the gap does not increase (see **Figure 5a**). This means that this contradiction does not arise with the Kogelnik model but rather is due to a simplified explication of the oscillation amplitude conditions when a laser medium with a bandgap is involved in forming the amplification spectra of the laser medium.

The authors of Ref. [26] predicted that the ban of amplification at the boundaries of the gap and the efficient amplification at the edges of the gap depend on an efficient lifetime of photons in the pointed spectral parts. The lifetime of photons confined by a bandgap indefinitely increases, meaning that they virtually escape. In contrast, photons on the edges of the bandgap increase their lifetime once and again, which means that they amplify by being in the gain region. Among our most critical physical results, we emphasize the direct observation of photons confined by a chiral grating and the bandgap shown in the fluorescence spectra in **Figures 3** and **4** for the laser medium without a bandgap and **Figures 6** and **7** for the bandgap.

Because the presented review bears evident tracks of the author's academic experience, monumental achievements LC in region display technology and optical waveguide communication have not received much attention. The technology and principles of these achievements are naturally represented in many textbooks, and manuals are required for their application. That is why we can try to comfort the readers and us with some references to bright virtual publications of the modern times.

To date, many authors have reported the successful performance of "diode-pumped a dye-doped LC laser." The new reference Brown C.M. with coauthor reported it in https://doi.org/10.1016/j.optlastec.2021.107080. When this result will get reproducible, many of the following major problems will be solved: polychromatic liquid crystal laser arrays for display applications [28], flat-panel laser displays based on liquid crystal microlaser arrays [29], and 3D laser displays based on circularly polarized lasing from cholesteric liquid crystal arrays [30].

In addition to improved display technology, reliable DFB chiral LC lasers are considered potentially valid for specific tasks such as on-demand polarization control and switching of lasing direction in reconfigurable chiral photonic crystals [31, 32].

As before, critical tasks for DFB chiral LC lasers remain in the region of frequency tuning. The authors of Ref. [33] proposed a proven electrical change in the refractive index for lasing in defective modes of the photonic LC.

Author details

Eugene A. Tikhonov* and Igor P. Ilchishin
Institute of Physics, National Academy of Sciences, Prospect of Science, Kyiv, Ukraine

*Address all correspondence to: etikhon@gmail.com

References

[1] Sorokin P, Lancard J. Stimulated emission observed from an organic dye ChAL-phthalocyanine. IBM Journal of Research and Development. 1966;**10**(2):162-163. DOI: 10.1147/rd.102.0162

[2] Schafer F, Schmidt W, Marth K. New dye lasers covering the visible spectrum. Physics Letters A. 1967;**24**(5):280-281. DOI: 10.1016/0375-9601(67)90437-9

[3] Tikhonov E, Shpak M. Stimulated light radiation by liquid solutions of polymethine dyes. Ukrainian Journal of Physics. 1967;**12**(9):2077-2078

[4] Stepanov B, Rubinov A. Stimulated optical emission of organic dye solutions. Journal of Applied Spectroscopy. 1967;**7**(4):345-354

[5] Ilchishin I, Tikhonov E, Shpak M, Doroshkin A. Stimulated emission lasing by organic dyes in a nematic liquid crystal. JETP Letters. 1976;**24**(6):336-339

[6] Tikhonov E, Shpak M. Superluminescence light emission by organic dye solutions. Ukraine Physical Journal. 1970;**15**(2):341-343

[7] Kogelnik H, Shank C. Coupled wave theory of distributed feedback lasers. Journal of Applied Physics. 1972;**3**(5):2327-2335. DOI: 10.1063/1.1661499

[8] Tikhonov E, Ilchishin I. The distributed feedback laser based on doped nematic liquid crystal. Ukrainian Journal of Physics. 2009;**4**(1):28-32

[9] Zabello E, Tikhonov E. One-mode oscillation of a dye laser with distributed feedback. Technical Physics Letters. 1979;**5**(17):1054-1056

[10] Bezrodny V, Zabello E, Tikhonov E. Picosecond distributed feedback. Quantum Electronics (FSU). 1984;**11**(12):2439-2442. DOI: 10.1070/QE1984v014n12ABEH006235

[11] Vashchuk V, Gorot K, Zabello E, Tikhonov E, Shpak M. New dye laser scheme with dynamic DFB. Quantum Electronics (FSU). 1978;**5**(7):144-152. DOI: 10.1070/QE1978v008n07ABEH010436

[12] Chandra S, Takeuchi N, Hartman S. Prism dye laser. Applied Physics Letters. 1972;**21**(4):144-146

[13] Bondar M, Przhonska O, Tikhonov E, Fedotkina N. Thermal optics of dye-doped elastomers. Journal of Technical Physics. 1986;**56**(5):878-883

[14] Ilchishin I, Tikhonov E, Tishchenko B, Shpak M. Frequency tunable dye laser with cholesteric LC Bragg mirror. Quantum Electronics (FSU). 1978;**5**(12):2637-2640. DOI: 10.1070/QE1978 v008n12ABEH011397

[15] DeVries H. Rotatory power and other optical properties of certain liquid crystals. Acta Crystallographica. 1951;**4**:219-226. DOI: 10.1107/S0365110X51000751

[16] Ilchishin I, Tikhonov E, Tishchenko V, Shpak M. Generation of a tunable radiation by impurity cholesteric liquid crystals. JETP Letters. 1980;**32**(1):24-27. DOI: 10.13140/RG.2.1.2672.4565

[17] Ilchishin I, Kleopov A, Tikhonov E, Shpak M. Stimulated tunable radiation in an impurity cholesteric liquid crystal. Bulletin of the Academy of Sciences of the USSR. Physical series. 1981;**45**:1376-1383

[18] Belotsky E, Ilchishin I, Lev B, Tomchuk P, Shpak M. Investigation of the optical properties of liquid crystals with induced gyrotropy by laser spectroscopy. Ukrainian Journal of Physics. 1991;**36**(2):239-249

[19] Ilchshin I, Tikhonov E, Mikityuk T. Narrowing the oscillation spectra of a cholesteric liquid crystal laser. Molecular Crystals and Liquid Crystals. 2019;**670**(1):112-118. DOI: 10.1080/15421406.2018.1542071

[20] Ilchishin I, Tikhonov E. Dye-doped cholesteric lasers: Distributed feedback and photonic bandgap lasing models. Progress in Quantum Electronics. 2015;**41**:2-22

[21] Yablonovich E. Inhibited spontaneous emission in solid-state physics and electronics. Physical Review Letters. 1987;**58**:2059-2062. DOI: 10.1103/PhysRevLett.58.2059

[22] Ilchishin I, Tikhonov E, Fedoryako A, Shpak M. Structure harmonic distortion of nematic with induced hyrotropy and lasing. Ukrainian Journal of Physics. 1988;**10**:1492-1494

[23] Ilchishin I, Tikhonov E, Tolmachev A, Fedoryako A, Shpak M. Harmonic distortion of induced helical structure of nematic LC detected by DFB lasing. Molecular Crystals and Liquid Crystals. 1990;**191**:351-355. DOI: 10.1080/00268949008038617

[24] Ilchyshyn I, Tikhonov E. How laser physics brought optics to the world of photonic crystals. Ukrainian Journal of Physics. 2020;**4**:327-334. DOI: 10.15407/ujpe65.4.327

[25] Kopp V, Fan B, Vithana H, Genak A. Low-threshold lasing at the edge of a photonic stop band in cholesteric liquid crystals. Optical Letters. 1998;**23**:1707-1709. DOI: 10.1364/OL.23.001707

[26] Douling J, Scalora M, Broemer M, Bowden C. The photonic band edge laser: A new approach to gain enhancement. Journal of Applied Physics. 1994;**75**:1896-1899. DOI: 0.1063/1.356336

[27] Belyakov V, Semenov S. Optical edge modes in photonic liquid crystals. Journal of Experimental and Theoretical Physics. 2009;**109**:687-690. DOI: 10.1134/S106377610910015X

[28] Morris S, Hands F, Findeisen-Tandel S, Cole R, Wilkinson T, Coles H. Polychromatic liquid crystal laser arrays toward display applications. Optics Express. 2008;**16**:18827-18837. DOI: 10.1364/OE.16.018827

[29] Fa X, Yong L, et al. Flat-panel laser displays based on liquid crystal microlaser arrays. CCS Chemistry. 2020;**2**:369-375. DOI: 10.31635/ccschem.020.202000162

[30] Zhan X, Xu FF, Zhou Z, Yan Y, Yao J. 3D laser displays based on circularly polarized lasing from cholesteric liquid crystal arrays. Wiley Online Library Advanced Materials. 2021;**33**(37):2104418

[31] Taimoor A, Elston S, Jia-De L, Morris S. On-demand polarization controllable liquid crystal laser. Advanced Materials Technology. 2022;**8**(3):2200674. DOI: 10.1002/admt.202200674

[32] Feng TM, Wang YF, Tseng HY, Li CC. Switching of lasing direction in reconfigurable chiral photonic crystals. Optical Materials. 2023;**13**:1555-1562. DOI: 10.1364/OME.491580

[33] Sung G-F, Chiu S-Y, Chang Y-C, Liou Y-C, Yeh CP, Lee W. Electrically tunable defect-mode wavelengths in a liquid-crystal-in-cavity hybrid structure in the near-infrared range. Materials. 2023;**16**:3229. DOI: 10.3390/ma16083229

Chapter 7

Red-Emitting Fluorophores with Tailored Properties for Microscopy and Nanoscopy

Kirill Kolmakov, Massimiliano Lucidi and Alexey V. Nizovtsev

Abstract

Extended series of red-emitting rhodamine dyes were synthesized and tested. This revealed the real factors determining the quality of STED (stimulated emission depletion) imaging, dye photostability in general, and stability of the corresponding reactive labels. Previously overlooked and underestimated observations are now being highlighted and discussed. So are some very important syntheses-related aspects. The studies were also expanded on far-red emitting oxazine and silicon rhodamine dyes. An intriguing discovery emerged when certain fluorophores, which were originally designed for STED, demonstrated an ability to discern between viable and dead bacterial cells. Recent publications by *Lucidi* and co-workers showed that certain positively charged dyes allow imaging of bacterial membranes using both conventional techniques, for example, confocal laser scanning microscopy (CLSM), and STED. This opens avenues for investigating membrane biogenesis and diagnostics.

Keywords: dyes, fluorophores, microscopy, imaging, photostability, red-emitting, biology, STED, rhodamine, oxazine, bacteriology, bacterial membrane, LIVE/DEAD

1. Introduction

Far-field optical microscopy in its modern applications such as STED (stimulated emission depletion) [1–4], PALM (photoactivation localization microscopy) [5–7], STORM (stochastic reconstruction microscopy) [8], and GSDIM (ground state depletion with individual molecular return) [9] poses very strict and often contradictory requirements on fluorescent markers involved [10, 11]. Among those, there is a large quantum yield of fluorescence, a high extinction coefficient, high photostability, and good solubility in water. No less important is a proper reactive group for conjugation to biological substrates [4, 11]. The labels (usually dye active esters) need to be stable upon storage in a freezer in the neat state and as stock solutions. Indeed, all these features seem to be hardly met by a single substance. However, in the course of our targeted research work, we showed that it is possible, at least by a dye scaffold with multiple variable positions. One very important advantage of rhodamine dyes as a class is the unique combinatorial amenability, as we call it [11]. It means an opportunity to combine and vary diverse functional groups and linkers. As a result, one can obtain fluorophores with tailored properties, staying within the same dye family. Here, we are trying to summarize our studies on this subject. We will tell how one can attach the required moieties to different sites of a rhodamine dye scaffold and take advantage of this.

2. Dye design and applications

2.1 Design of photostable and hydrophilic red-emitting STED fluorophores available as reactive labels with improved stability

Red-emitting rhodamines with absorbance/emission maxima of around 635/655 nm are now quite numerous, as they have been known for more than a decade [10–12]. The first dye of this family suitable for STED microscopy was synthesized and explored by *Kolmakov* and co-workers. It is known as KK 114 (see **Figure 1** and Ref. [11]) and was supposed to be a better and cheaper alternative for the dye ATTO 647 N. The latter, in turn, was the first dye purposefully designed for STED microscopy by ATTO-TEC GmbH [13]. For this application, photostability and, especially, stability toward photo-induced reactive oxygen species were crucial. Its photophysical properties were indeed greatly improved compared to those of previously known labels, for example, Alexa Fluor 633. However, despite its merits, ATTO 647 N is too lipophilic (see **Figure 4**, right) and, as it is, leaves no opportunity for further chemical modification. In aqueous media, the dye sticks to glass and, with antibodies, often gives unspecific labeling, strong background, and aggregation. So, a better candidate needed to have polar functional groups. As the most promising alternative dye scaffold, a fluorinated red-emitting rhodamine was chosen, which had been already known from patent literature (see **Figure 1** and Ref. [14]).

The new rhodamine dye KK 114 (later known as Abberior STAR RED [15]) soon gained recognition due to its excellent STED imaging performance. It showed a very high brightness, high signal-to-noise ratio, and high resistance to bleaching, accompanied by low background and low unspecific labeling [9, 11, 12]. The image quality with certain substrates, particularly, mitochondria organelles [16, 17], is unparalleled. Importantly, with this fluorophore, one can even operate at "higher" STED beams (i.e., 800 nm instead of 775) [18], which are, by definition, less harmful for samples given very high beam powers. Later, phosphorylated, hydroxylated and photostabilized fluorophores were prepared (see **Figures 1–5**). Apart from other

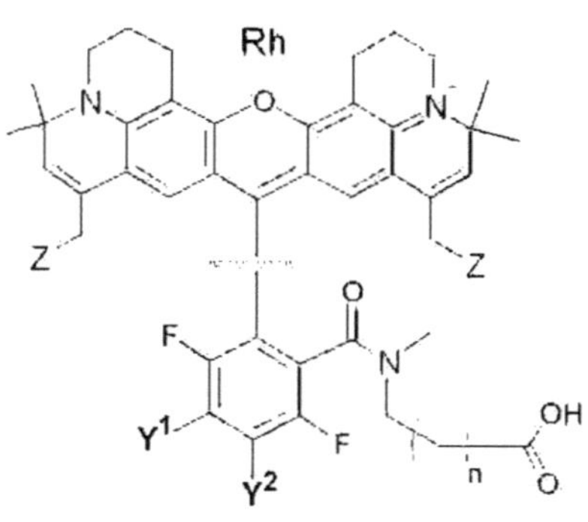

KK 1517	Y^1, Y^2 = F, Z = H, n = 2
KK 1103	Y^1 = $S(CH_2)_2SO_3H$, Y^2 = F, Z = H, n = 2
KK 9046	Y^1, Y^2 = F, Z = OH, n = 2
KK 1115*	Y^1 = $S(CH_2)_2SO_3H$, Y^2 = F. Z = OH. n = 2
KK 114	Y^1, Y^2 = F. Z = SO_3H, n = 1
KK 114L	Y^1, Y^2 = F, Z = SO_3H, n = 2
KK 114S	Y^1, Y^2 = $S(CH_2)_2SO_3H$, Z = SO_3H, n = 1
KK 1550	Y^1, Y^2 = F, Z = $OP(O)(OH)_2$, n = 2
KK 1555	Y^1 = $SCH_2(4\text{-}C_6H_4)NO_2$, Y^2 = F, n = 2, Z = SO_3H
KK 1556	Y^1. Y^2 = $SCH_2(4\text{-}C_6H_4)NO_2$, n = 2, Z = SO_3H

Figure 1.
*Dye scaffold and a family of red-emitting rhodamines, which were screened as fluorophores for STED microscopy. For details and syntheses see Refs. [9, 11]. * Compound KK 1115 is also known as Abberior STAR 635 [11, 12]. Rh denotes the dye core (see also* ***Figure 2****), while Y and Z are variable functional groups.*

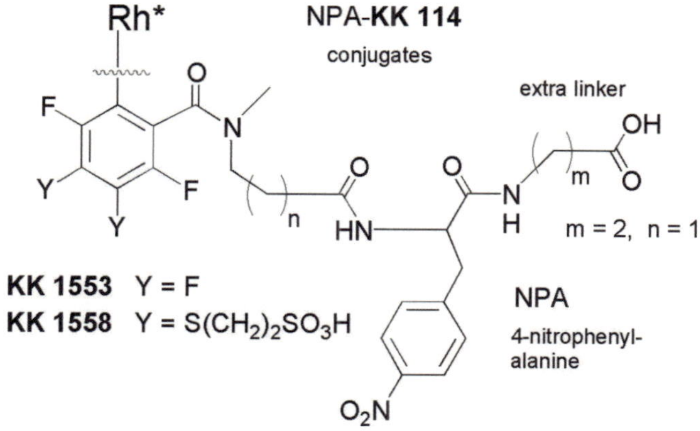

Figure 2.
*Red-emitting rhodamine KK 114 modified with an intramolecular stabilizer (NPA) and optional charged groups. *Rh denotes the core of the red-emitting rhodamine scaffold according to **Figure 1**.*

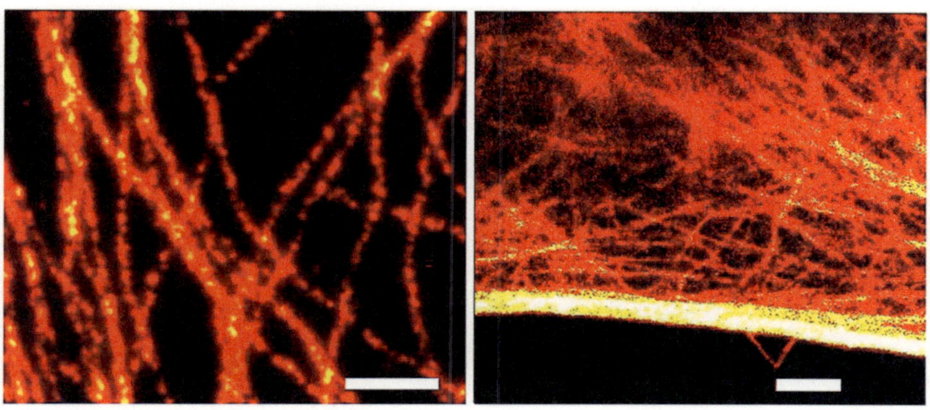

Figure 3.
Exemplary STED images of tubulin (left) and actin (right) labeled with Abberior STAR 635. For imaging and labeling conditions see Refs. [11, 12]. Scale bars: 1 μm; resolution ~40 nm.

benefits, the analogs of KK114 demonstrate the so-called "good application tolerance": they are far less sensitive to the changes in illumination/imaging conditions and other setup settings, developing fewer re-excitation artifacts. Here, the advantages of KK 114 compensate for the fact that it is not as suited to the far-red spectral area as, for example, Si-rhodamines [11, 18]. Last but not the least, the synthesis of KK 114, its derivatives, and building blocks is not very difficult. In fact, one needs no metalorganic chemistry, no dangerous chemicals, and no tedious separations either. All the chromatographic separations can be done using conventional hand-made columns. On the contrary, the synthesis of ATTO 647 N takes more than 11 steps and utilizes much more sophisticated precursors and recipes [13]. Finally, the dye KK 114, even taken as it is, can be further functionalized *via* aromatic nucleophilic substitution in the *meso*-phenyl ring [11, 19]. Particularly, as we shown in Ref. [11], it is possible to directly decorate this dye (or its precursors) with extra polar ionizable groups (e.g., $HO_3SCH_2CH_2S$-) or intramolecular stabilizers (e.g., 4-NO_2PhCH_2S-). Regarding the stability of the active esters, usually NHS (*N*-hydroxysuccinimidyl)

Figure 4.
Left: Red-emitting rhodamine dye STAR635 with a blue color and intense red fluorescence in aqueous solution under visible light. Right: Distribution of Abberior STAR635 (1) and ATTO 647 N (2) between dichloromethane (lower phase) and water at pH 8.5 (labeling buffer; upper phase). The dye ATTO 647 N is so lipophilic that it remains exclusively in the lower organic phase.

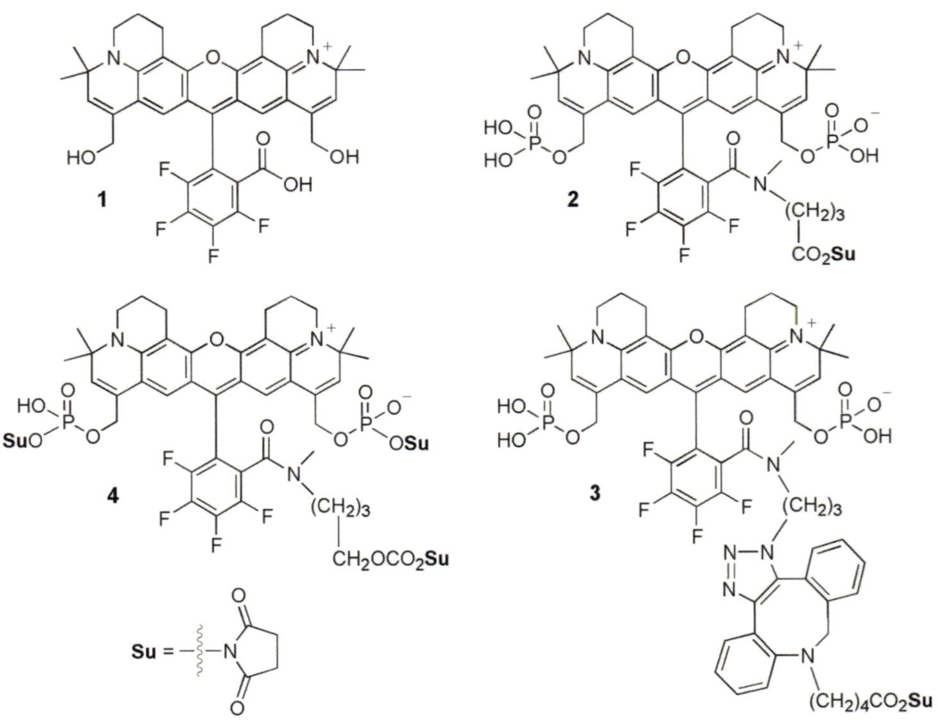

Figure 5.
*Synthesis of phosphorylated red-emitting rhodamines and their NHS esters. Compound **1** is the precursor for phosphorylation, and compound **2** is the desired NHS ester, which presents selectivity challenges during its synthesis. Compounds **3** and **4** represent approaches to solve these issues.*

or TFP (tetrafluorophenyl) esters, we made three important observations: (1) the linker size improves the stability, especially after certain critical length is exceeded. This can be attributed to the influence of the bulky chromophore residue. Particularly, antibody labeling proved unfeasible with the non-sulfonated analog of KK 114 (see Ref. [9]), so no images were recorded for comparative experiments. This instability was mistakenly attributed to the positive net charge of its molecule. Also, it proved impossible to prepare active esters from the 4-nitro-L-phenylalanine-modified dye KK 114 (see Sections 2.3 and 2.4) unless the alanine terminus was modified. We found that a simple elongation of the linker group even by one methylene unit makes the active esters in all dyes of the family, including KK 114, much more stable, easier to handle and to ship. This applies to mixed NHS carbonates of the rhodamines, as it was shown, for example, in Ref. [20] (see also compound **4** in **Figure 5**) and for fluorophores of other families (see **Figures 6** and 7), as well; (2) a large negative net charge of a label decreases the stability of its active esters, as we observed in a comparative study, which involved labels with extra SO_3H groups [11]; (3) chromatographic isolation of the NHS esters should be avoided where possible. All derivatives of dyes with multiple polar groups require aqueous or alcoholic media as the mobile phase, where their stability is low, especially upon concentrating. Particularly, rhodamine dyes of this family can react with methanol to give either Me-esters or, in basic conditions, compounds where one aromatic fluorine atom is exchanged for MeO or OH [9, 11, 19]. In view of all these, it is crucial to find such a recipe that easily brings the NHS esterification to completion. Having achieved that, one removes the excess of the reagents (small molecules and salts) by simple extraction or sonication of the solids in proper inert solvents [11, 21]. Among the NHS ester labels with the best stability, we should, first of all, mention STAR 635 (also called KK 1115 in the references above, with a net charge (z) of −1). It was purposefully developed as a label with improved stability upon handling and storage.

Particularly, STAR 635-NHS withstands aqueous workup in neutral or slightly acidic media and, therefore, can be extracted with proper organic solvents that are easy to evaporate. This is followed by drying and simple filtration. These procedures quite well remove salts and silica gel dust, which is, at times, a big problem [11]. Preparative chromatography (hand-made columns or HPLC) at pH ≤ 7 followed by concentrating at room temperature under reduced pressure makes no harm either. The NHS esters of KK 114 L-NHS, KK 1553-NHS, KK 1555-NHS (all three have z = −1), and especially KK 1517-NHS (z = +1, with no negatively charged groups)

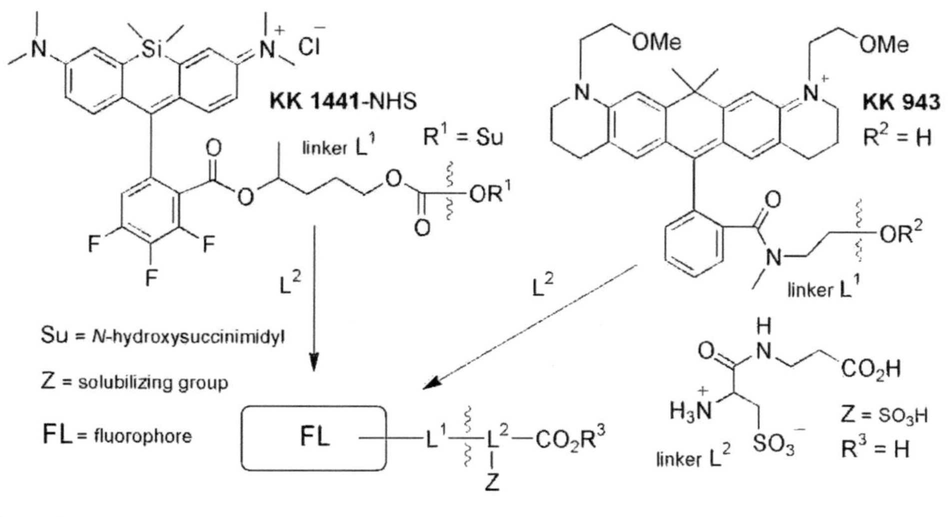

Figure 6.
Exemplary post-synthetic modification that affects the net charge and makes the fluorophores hydrophilic. A general formula of the dyes modified with custom-made branched linkers L^2 and auxiliary (solubilizing) groups.

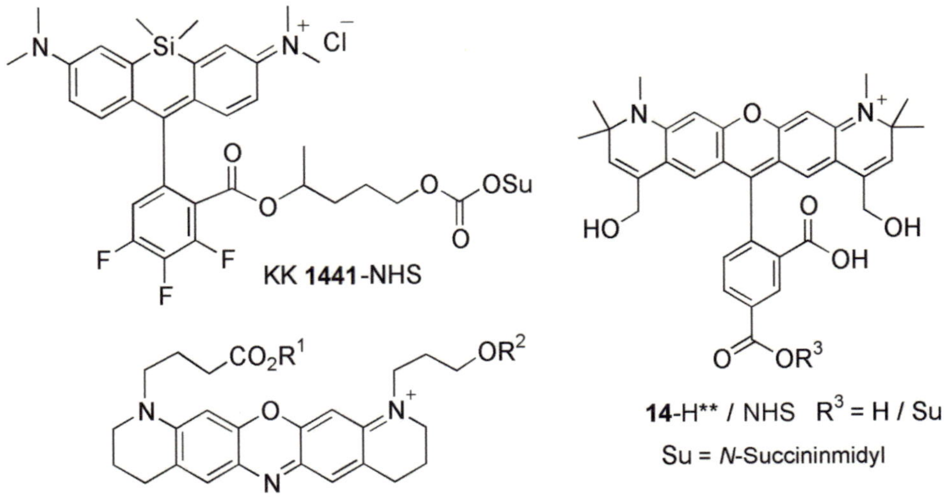

2-H / **2**-NHS R^1 = H / Su, R^2 = P(O)(OH)$_2$
KK 1905* / **1905**-NHS R^1 = Et, R^2 = H / C(O)OSu

Figure 7.
*Fluorophores from other families for multi-color imaging shifted to the far-red (KK 1441, and 2-H, see Ref. [18]) and rhodamine shifted to the orange spectral region (14-H). * – KK 1905-NHS is a custom-made non-phosphorylated oxazine label with a positive net charge for bacterial imaging [21]. ** – for synthesis, properties, and STED imaging see Ref. [20] and* ***Figure 13***.

Figure 8.
Left: dye KK 114 as a fluffy solid after freeze-drying. Note the red glittering under daylight. Right: evaporation of the KK 114-NHS ester aliquots on small scale (1 mg of the solid) using a vacuum-inert gas manifold [11].

are sufficiently stable too [11]. Regarding our far-red emitting labels, the stability of KK 1905-NHS (z = +1) proved to be surprisingly good (see **Figure 7** for structure and [21] for preparation and properties). Straightforward and reliable recipes for the preparation of the NHS esters can be found in the references above (see also **Figure 8**). Finally, imaging experiments with various labels showed that a dye linker does not need to be as short as possible, as was mistakenly assumed earlier [4]. Particularly, in the sections below, one will see the dyes with bulky extra moieties and with longer and/or branched linkers. They all performed very well.

2.2 Phosphorylated dyes as an alternative to the conventional sulfonated compounds

The introduction of sulfonic acid groups (sulfonate) to aromatic compounds, known for more than 150 years, drastically improves the solubility in water, which is important for many industrial dyestuffs, fluorophores for microscopy, and laser dyes [4, 11]. Having obtained the diol precursor **1** (**Figure 5**) by oxidation of the corresponding rhodamine fluorophore at the allylic sites of the parent rhodamine, we sought to modify the hydroxy (OH) groups so that the dye could get ionizable functional groups with acidic protons. Free OH groups offer a variety of opportunities for derivatization and, at the same time, remain unreacted while the carboxyl-containing linker is subjected to a peptide-type coupling. So, an alternative rhodamine dye candidate, which is modified at the OH sites, must be an ester of a strong mineral acid. Esters of mineral acids raise hydrolytic stability concerns in multistep syntheses and while labeling. Therefore, one should exclude nitric and sulfonic acid esters as candidates. Phosphate was thus chosen as an alternative to the conventional sulfonate. Indeed, this acid has a medium strength, an increased number of acidic protons relative to sulfuric acid, and most importantly, its primary and secondary alkyl esters are hydrolytically stable in basic media. So, one can increase the net charge of a dye as a whole, which is very important for electrokinetic analyses. Also, one can call the phosphate group "biologically friendly", as it is very abundant in living nature. However, before our first studies on dye phosphorylation were published in 2012, no attempts had been known to attach phosphoric acid (phosphate) residues to fluorescent dyes [22]. Phosphate groups ($-OP(O)(OH)_2$), despite these advantages, pose selectivity concerns while preparing reactive labels [20]. These groups do react with most coupling reagents to a noticeable extent to form P-NHS esters (see compound **4** in **Figure 5**). Unexpectedly, the latter proved inert toward amines. Despite the praiseworthy attempts, an efficient and reasonable way how to selectively activate a carboxyl group ($-CO_2H$) in the presence of a phosphate was not found. Remarkable is the approach that involved a dye with free phosphate groups, an azide, and a functionalized alkyne as a counterpart. The dye azide was subjected to a metal-free click reaction with an alkyne that had a pre-synthesized amino reactive (NHS) site. It was a linker-containing dibenzocyclooctyne-NHS ester (DBCO-NHS). Obviously, such a bulky and lipophilic building block, which comprises more than 30% of the molecular mass of the entire conjugate (compound **3**), does not seem to be a reasonable solution. On the other hand, metal-catalyzed click reactions on smaller synthons proved unsuitable due to strong complexation [20]. Remarkably, the real (remeasured) fluorescence quantum yields Φ_f of the phosphorylated dye (compound **2** in **Figure 5** as acid; see also the dye KK 1550 in **Figure 1**) did not exceed those of "the benchmark dye" KK 114 with conventional SO_3H groups. Certain increases in the quantum yield were, however, observed [11], but only at pH above 9, which is unacceptable for most biology-related applications. So, summarizing, phosphate groups as auxiliary groups for microscopy-related fluorophores proved to be not worthwhile, yet they did involve certain interesting chemistry by their preparation. The phosphorylated fluorophores, however, proved to be very useful for electrokinetic-related applications, particularly for high-throughput carbohydrate analyses, as was reasonably assumed in [22] and shown in applications [23, 24].

2.3 Post-synthetic (last minute) modifications of the fluorophores

In the course of our research work, we very often made use of the combinatorial possibilities of the dye scaffold depicted in **Figure 1**. Particularly, we performed the so-called "post-synthetic" or "last minute" modifications. This approach was first realized by *Romieu* and co-workers in 2008 [25–30]. It is usually the linking group of a dye (as shown in **Figure 6**) that allows such transformations. High-yielding peptide-type coupling reactions were utilized. One can either modify the linker step by step or attach a pre-synthesized building block containing the required auxiliary (i.e., ionizable and/or solubilizing) groups. We had shown already how one could attach a required moiety to different sites of our specific dye scaffold. It can be done even in the final steps of the synthesis. Moreover, intramolecular photostabilizers (see Section 2.4 and **Figure 2**) can be installed following this straightforward methodology ("last minute" modifications). Luckily, sulfonic acid groups (sulfonate, $-SO_3H$), the classical dyestuff solubilizing moieties, are unaffected neither in amidation reactions nor in the preparation of active esters [11, 27, 28, 30]. Phosphate groups, on the other hand, pose selectivity problems, as it was mentioned above and in Ref. [20].

2.4 Fluorophores with intramolecular photostabilization

Self-healing dyes with photostabilizing moieties represent an efficient solution for the fatigue/bleaching problem [11, 31–36] (see **Figure 2**). However, too many questions remained unanswered. In our recent study, we tried to address them [11]. We wanted to establish to what extent the photostability can be further improved, and what it would be traded for. Particularly, the effect of intramolecular stabilizers on rhodamines, which are alone much more stable than cyanines, still remained unclear. The basic photophysical properties of the individual rhodamine conjugates containing photostabilizing groups (PSG), which were originally called triplet state quenchers (TSQ). Fluorescence quantum yields Φ_f (fl. QY), fluorescence lifetimes τ (FLT), and the photostability of the individual conjugates were thoroughly measured. These data were compared to their STED performance in the immobilized state and in bulk aqueous solutions, as seen in **Figures 9** and **10** (which contain parts reproduced from Ref. [11]). The authors took a closer look at the photostability and STED performance of the conjugate "NPA-KK 114" (dye KK 114 decorated with a 4-nitro-L-phenylalanine moiety – NPA (the conjugate was also mentioned as KK 1553, see **Figure 2** for structure and **Figure 10** for STED image series). Also, 4-nitrobenzyl thiol (NBT) was shown to be a good alternative for NPA (4-nitrophenylalanine) or NPAA (4-nitrophenylacetic acid) as a protective moiety [11, 36, 37]. As examples, one can see compounds KK 1555 and KK 1556 in **Figure 1**.

Experiments on aerobic photolysis in bulk dye solutions showed the following: the unmodified rhodamines always lose their extinction consistently, while the dyes are being split into smaller colorless molecules. At the same time, the PSG-modified dyes paradoxically increase their fluorescence signal in the beginning of the photolysis [11] and survive much longer than the parent dyes, as seen in **Figures 9** and **10**. However, the extinction of the modified dyes does constantly decrease, yet not as fast as it does in the unmodified compounds. The paradoxical rise in the fluorescence signal in the PSG-modified dyes under illumination can be explained by assuming a dye-mediated

Figure 9.
Photolysis in solution upon stirring in air: Reference dye KK 114 and an "everlasting dye" - its conjugate with an intramolecular stabilizer (NPA). The blue color fades, as seen in daylight, and so does the initial red color of the fluorescence, as seen under UV illumination (365 nm) in the dark. Note the drastic spectral shift of fluorescence (bluing). The unmodified dye is thus almost completely "burned out" under illumination with a powerful 400-wt. halogen lamp [11].

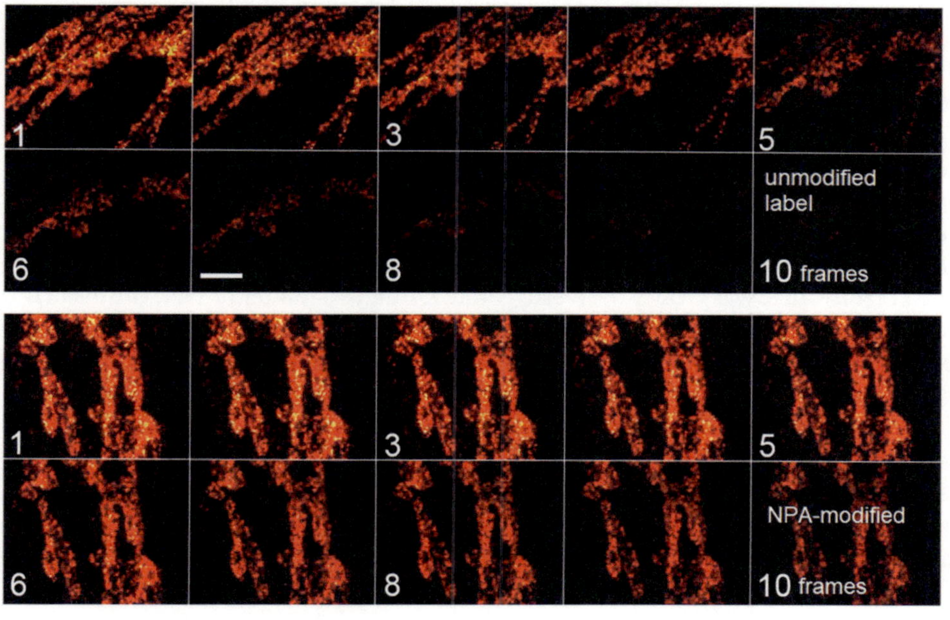

Figure 10.
The reference dye KK 114 L (top) and its NPA-modified derivative KK 1553 (bottom) in STED imaging. Presented are a series of ten STED frames of a mitochondrial inner membrane Mic60 with secondary antibodies custom-labeled with the fluorophores. Scale bar: 1 μm. The unmodified label fades completely after taking ten STED frames [11].

photo-cleavage of the photostabilizing group (4-nitrobenzyl). In the course of the cleavage, the "free dye" with a much higher Φ_f is released (see Ref. [11] and Table 1 therein), so the resulting fluorescence signal becomes higher than it was at the beginning. We assume that 4-nitrobenzyl, as a part of NPA and NBT moieties, forms radicals and/or quinoidal structures, analogously to 2-nitrobenzyl, a conventional photolabile protecting group (PPG) [38]. This allows the 4-nitrobenzyl group to relieve the excitation energy from the fluorophore and even to play a sacrificial role. In a FRET pair, the FLT of the donor becomes shorter, as a favorable channel of energy transfer from the excited (S1) state appears [39]. Here the transfer is, however, non-radiative. Also, neither the NPA nor NBT groups absorb visible light. This is completely different from the way how the PPGs work. Summarizing, one can say that the NPA- or NBT-modified dyes can fulfil the demand for stable fluorophores for prolonged microscopical observations. It is noteworthy that the attempts to obtain an NHS ester from the conjugate "NPA-KK 114" failed until an extra linker was installed (see **Figure 2**). Finally, a praiseworthy idea to improve the photostability by means of a partial deuteration did not work for KK 114 [9]. One could assume that, back then, a wrong site of the fluorophore was modified. However, a decade later, a noticeable increase [10] in photostability was observed when fluorophores were deuterated at their real weak points – terminal *N*-alkyl sites.

2.5 Do not count the photons, make the photons count!

In addition to the dye design, we tried to discover the real determining factors in the STED microscopy performance of a dye family. In view of our recently presented data, one had to reconsider some well-established views on this topic [11]. Our image series showed that, contrary to the well-established opinion [9, 12], the photostability, not the fluorescence quantum yield of the fluorophore, is the real determining factor in STED imaging. Also, the presence of polar groups and the net charge are not as important as they seemed to be. So, the "race" for very high fluorescence quantum yields is not always justified. Immunolabeling and imaging with the "control dye," which has no extra polar groups at all, demonstrated no increase in unspecific labeling and no increase in the background either, as seen from the control STED images in [11] and in related supporting information. Note that unspecific labeling and background were previously attributed exclusively to the absence of polar functional groups [9, 12, 22]. Their significance was clearly over-exaggerated. But, as a matter of fact, the decoration of fluorophores with polar functional groups, hydroxy group (-OH) including, was not in vain. As we established [11], these groups do increase photostability, which proved to be the real deciding factor in STED imaging. It was shown that, even with a fluorophore whose Φ_f is as small as 14–17%, good-quality STED images can be taken, as seen in **Figure 11** (partly reproduced from Ref. [11]). They were recorded with a sufficient photon budget and with only a small loss in resolution and signal-to-noise ratio, compared to the reference dyes. Very importantly, one should keep in mind that, in immunolabeling one operates with antibodies whose molecular weight (MW) is usually around 150 kDa, whereas the MW of usual small-molecule fluorophores does not exceed 1 kDa. So, a label is obviously not big enough to influence the properties of an antibody conjugate as a whole, particularly its hydrophilicity and polarity. In other words, the antibodies that are usually used for immunolabeling are very hydrophilic and bulky. On top of that, the objects are of micrometer size and, in biology-related applications, are by definition never identical. Neither is the actual amount of a fluorophore in the site under the focus. Substance diffusion and heat dissipation in a viscous medium might be hindered. Finally, the objects can drift away from the focus. All these make it problematic to compare and quantitatively estimate the performance of dyes, especially without statistics. One should definitely hold himself from making far-going conclusions on STED performance of dyes without having at least some statistics.

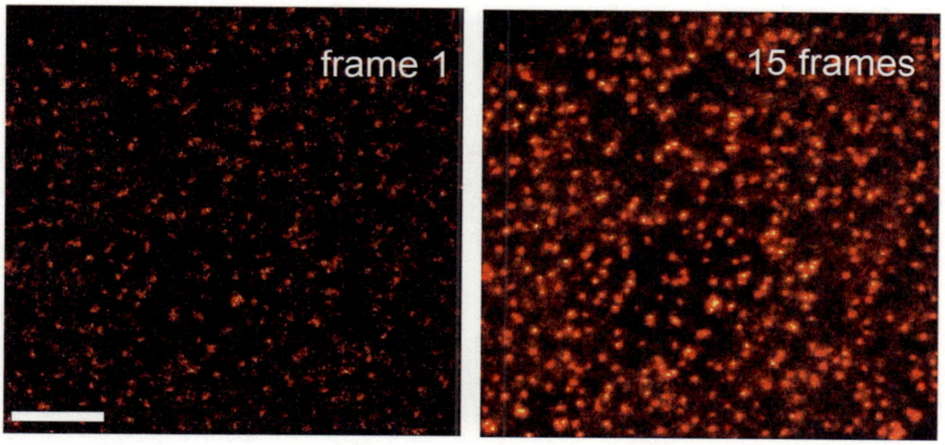

Figure 11.
Exemplary STED images with a photostabilized model dye KK 1556 having a fluorescence quantum yield (Φ_f) of only 14%. Presented are the first frame and an add-up of 15 frames of an immunolabeled nuclear pore complex. See Ref. [11] for imaging details and conditions. Scale bar: 1 μm.

Immunolabeling involves two steps, so the image quality and an unspecific background are very often just a matter of good fortune and/or tidiness. There is, however, one obvious thing that one could state for sure: in the ideal case, one needs a good batch of antibodies, a good label, a good experimentalist, and favorable imaging settings.

2.6 Additional spectrally shifted fluorophores for multicolor imaging

For multi-color imaging, a spectral separation of different labels in two or more excitation or acquisition channels is essential. Alternatively, time-resolved imaging can be performed, where fluorophores with different excited state lifetimes (FLT) are employed [20, 40]. Neither spectral nor FLT separation of two labels is usually not possible within a single dye scaffold where simple functional group transformations are performed. To that end, one needed to switch to another chromophore family. As such, custom-made far-red oxazine and Si-rhodamine dyes (see **Figure 7** for structures and **Figure 12** for images) were studied [18], which allow both options (a FLT difference and a spectral shift). For imaging in a two-color STED with spectrally separated labels, we used compounds STAR RED and 14-H (another red-emitting rhodamine [20] with solubilizing OH groups, see **Figure 7** for structure), whose absorbance and fluorescence maxima are 636/659 and 579/609 nm, respectively. Direct color separation in two observation channels with a negligible background signal was achieved with different cellular substrates, as seen in **Figure 13**.

A later work by *Winter* and co-authors [40] was a milestone for multi-color imaging: it illustrates the approach showing four-color nanoscopy of fixed and living cellular samples. In that work, a far-red emitting silicon rhodamine KK1441, one of the four dyes, (see **Figure 7** for structure) played an important role. The authors came up with "hyperSTED" – a simple and versatile STED instrument design for multicolor nanoscopy. This utilized only two laser sources. Particularly, only one excitation laser and a single STED beam are needed. By hyperspectral detection of four channels, separation of four different markers was accomplished. Discrimination of the markers by the differences in emission spectra allowed fast and simultaneous data acquisition [40]. To realize such an approach in practice, a four-color fixed cell sample was stained with a suitable

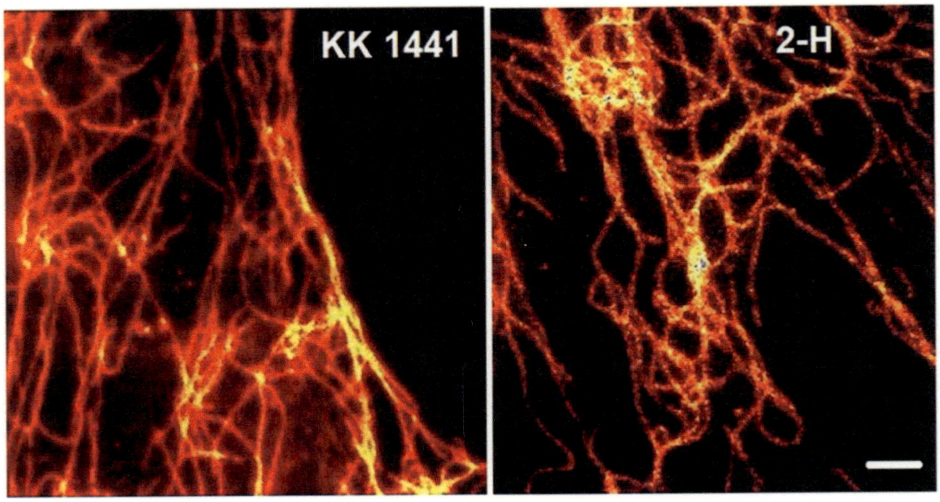

Figure 12.
*Exemplary STED images with a Si-rhodamine- (KK 1441) and a phosphorylated oxazine-containing (2-H) far-red emitting fluorophores [18]. Scale bar: 1 μm. See **Figure 7** for structures.*

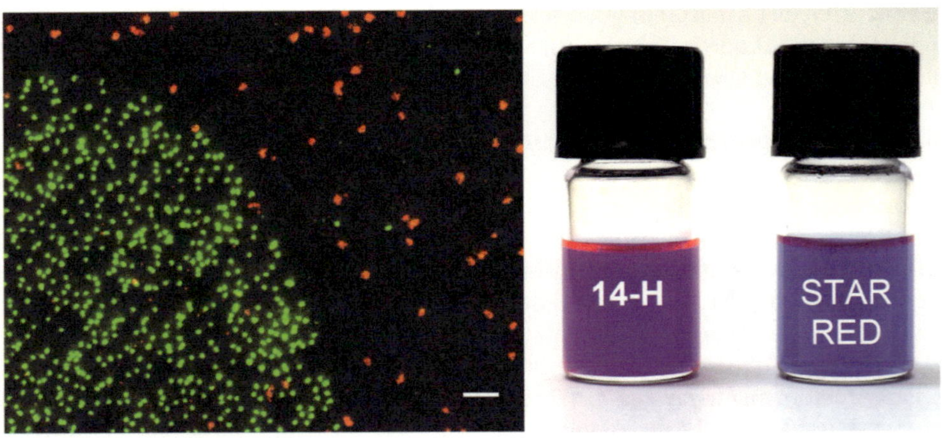

Figure 13.
*Two-color STED image with a pair of spectrally separated red-emitting fluorophores STAR RED (spectrally identical to KK 114, see **Figure 1** and [15, 20]) on PMP (in red) and 14-H on NUP 153 (in green, see **Figure 7** for structure). Spectral difference is clearly seen in solutions, as well. For details see Ref. [20]. Scale bar: 1 μm.*

combination of four of the tested dyes with the following abs./fl. maxima (nm): Atto594 (601/627) for peroxisomes, Abberior STAR 635P (633/659) for vimentin, KK1441 (661/679) for giantin and CF680R (680/701) for nuclear pores.

3. New bacteriology-related applications of red and far-red emitting dyes

3.1 The microbiologist palette: synthetic dyes and fluorescent proteins

The increasing use of fluorescence microscopes and CLSM caused a high demand for new dyes. Max Haitinger (1868–1946) introduced for the first time the term *'fluorochroming'* or fluorescent staining of specimens, and he developed some staining techniques for fluorescence microscopy [41]. Numerous advancements have emerged in the realm of fluorophore development after the groundbreaking discovery of the green fluorescent protein (GFP) in 1962 by Shimomura and colleagues [42, 43]. Despite the invaluable utility of GFP and their variants in tracking protein expression and localization within cells, they often exhibit a lower quantum yield compared to organic dyes [42]. Additionally, challenges such as fluorescent protein oligomerization, fluorescence attenuation during fixation, and moderate cytotoxicity upon overexpression persist [42]. Consequently, there is a growing demand for novel organic dyes, particularly, characterized by reduced steric bulk and accelerated labeling kinetics [44, 45]. These dyes are based on various molecular scaffolds including fluorescein, cyanine, oxazine, squaraine, BODIPY, xanthene, benzobisthiadiazole, and rhodamine [44]. For a more in-depth discussion on the use of these fluorophores in bacteriology, the reader is referred to the reviews [44–46]. The impact of fluorescent dyes in bacteriology further increased when some fluorophores were *ad hoc* generated to differently stain viable and dead bacteria, thus giving additional information on bacterial metabolic activity and/or membrane integrity under the exposition to environmental stressors (*e.g.*, antibiotic treatment). The predominantly employed combination of synthetic dyes in bacteriology involves SYTO 9 and propidium iodide (PI), commercially available in the LIVE/DEAD BacLight Bacterial Viability and Counting Kit (Thermo Fisher). The green fluorescing dye SYTO 9 penetrates inside a bacterial cell, and it is used to assess the total cell count, whereas the red fluorescing dye PI enters only cells with damaged membranes [47]. The emission properties of the stain mixture bound to DNA change due to the displacement of SYTO 9 by PI and quenching by fluorescence resonance energy transfer [48]. Increasing the presence of membrane discontinuities in cells facilitates the entry of PI, resulting in higher levels of SYTO 9 fluorescence quenching and emission of red fluorescence. Other techniques involve LIVE/DEAD staining, spanning from flow-cytometry [49], fluorimeter and other automatic fluorescence readers, confocal microscopy imaging [50], and fiber-based spectroscopic systems known as optrodes [51, 52].

3.2 The dark side of the microbiologist toolbox: the red and near-infrared emitting dyes

Red, far-red (FR), and near-infrared (NIR) emitting dyes offer significant benefits in bio-imaging within living systems. Their fluorescence in the longer wavelength range minimizes phototoxicity to biological components, enables deep tissue penetration, and reduces background interference from auto-fluorescent biomolecules [53]. In the last years, many efforts have been focused on the repurposing of fluorophores for bacterial imaging [44]. Luckily, among other candidates, the dye KK 114 (commercially available under the name Abberrior STAR RED [15, 21]) was tested as a contrast agent. It proved to perform very well, which was nicely demonstrated in the recent works on eukaryotic cells [21, 54–56]. This fluorophore is water-soluble, highly photostable, and negatively charged ($z = -1$ in the bound state). This makes its conjugates tentatively not permeable for eukaryotic cells. Surprisingly, despite its negative charge, this molecule demonstrated an ability to adhere to the membranes of various bacterial species, encompassing both Gram-positive and Gram-negative bacteria [54, 55]. Notably, the outer membrane of Gram-negative bacteria and the cytoplasmic membrane/cell wall of Gram-positive bacteria are decorated with lipopolysaccharide and teichoic acids, respectively, both making their outer shells negatively charged. This suggests that charge repulsion alone is insufficient to hinder the dye from binding to the membrane, implying the presence of other yet-to-be-identified mechanisms governing the interaction between the dye and bacterial macromolecules. Despite the unclear mechanism of action, KK 114 proved to be suitable for staining different members of the ESKAPE group. ESKAPE is the acronym for the group of *Enterococcus faecium*, *Staphylococcus aureus*, *Klebsiella pneumoniae*, *Acinetobacter baumannii*, *Pseudomonas aeruginosa* and *Enterobacter* species, which cause life-threatening hospital-acquired infections, mainly among critically ill and immunocompromised patients. These species are considered critical-priority pathogens, representing a global menace because of their high transmissibility, mortality, limited treatment options for the rapid diffusion of multidrug-resistant strains, and healthcare and community burden [57].

The capability to label bacterial membranes opens the vault to track the arrangements that occur during cell division. The inhibition of the bacterial division machinery (namely the divisome) gives a background for the generation of new antimicrobial compounds capable of interfering with the natural dynamics and functions of bacterial replication [58] with implications in ESKAPE containment, treatment, and eradication. Nevertheless, the imaging of bacterial membranes faces challenges due to the <10 nm dimensions of this structure, significantly below the maximum limit of spatial resolution achievable by modern confocal microscopes, which is approximately 200–250 nm [59]. To overcome this limitation, super-resolution approaches are required to achieve a suitable resolution for membrane imaging. STED microscopy is in fact an example of a CLSM-derived super-resolution technique. It is capable of pushing the resolution limits beyond the diffraction barrier [1–3, 59]; a thorough examination of super-resolution approaches is provided in the reviews by Refs. [46, 60]. Utilizing the remarkable photostability of KK 114, STED imaging was performed on different bacterial species (**Figure 14**), thus enabling the observation of bacterial membranes with a resolution one order of magnitude greater than that achievable with CLSM [21].

In STED imaging, a secondary KK 114 mechanism of action emerged: this molecule demonstrated a capability to penetrate bacterial cells with compromised membranes with a higher fluorescence emission (as indicated by the cyan arrows in **Figure 14**), suggesting a potential for conducting LIVE/DEAD staining. To that end, one should combine KK 114 with other compatible fluorophores [21, 54, 55]. Apart from its potential application in labeling division septa for exploring molecules to enhance our therapeutic options against ESKAPE pathogens, another interesting feature of KK114 was unveiled. It is also a powerful tool to perform rapid detection of colistin resistance in *A. baumannii* [55]. Colistin is a positively charged membrane-disrupting peptide used as a "last-resort" drug to treat infections caused by *A. baumannii* multidrug-resistant strains. This peptide interacts with the outer membrane, a distinct feature of Gram-negative bacteria, leading to the displacement of divalent cations that stabilize the outer membrane, thus causing derangement of the cell membrane, leakage of intracellular contents, and ultimately cell death [56]. A multicolor labeling strategy based on the combination of SYTO 9 and KK 114 allowed discrimination of colistin-resistant from colistin-sensitive *A. baumannii* strains [55]. In colistin-sensitive strains, exposure to colistin causes loss of membrane integrity that results in KK 114 intracellular penetration accompanied by an enhancement in FR fluorescence emission. Conversely, the membranes of colistin-resistant strains are impermeable to KK 114, with no detectable increase in FR emissions even at high colistin concentrations. Therefore, in colistin-sensitive strains, the ratio between

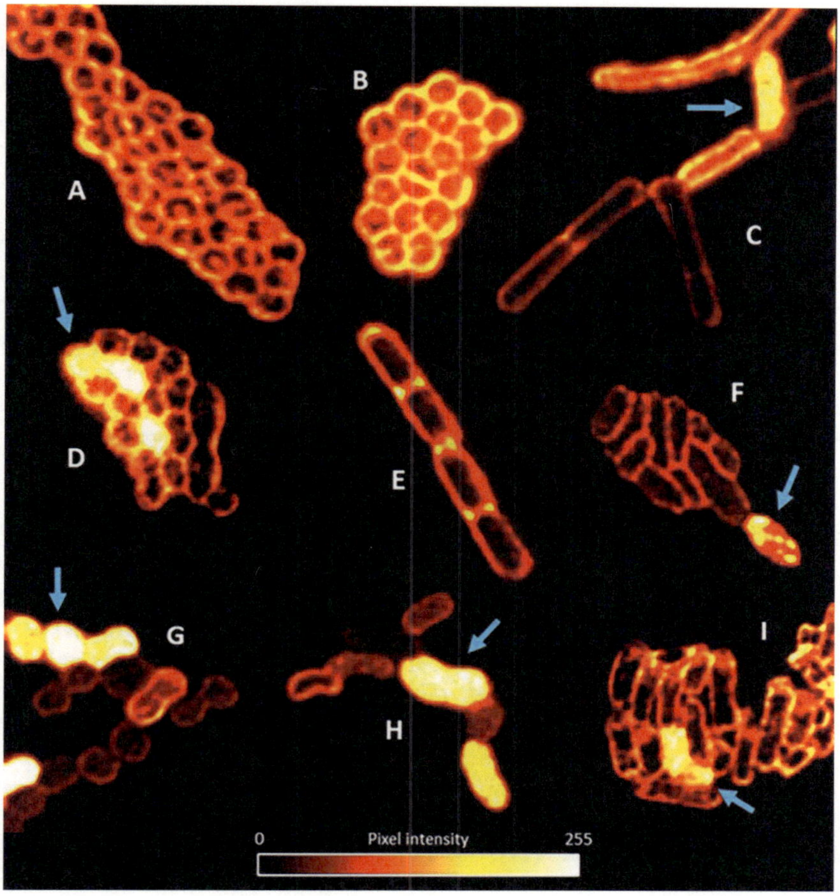

Figure 14.
STED imaging of different bacterial species stained with the dye KK 114: E. faecium *(A);* S. aureus *(B);* Bacillus subtilis *(C);* Streptococcus pyogenes *(D);* Escherichia coli *(E);* Achromobacter xylosoxidans *(F);* A. baumannii *ATCC 17978 (G);* A. baumannii *ATCC 19606^T (H); and* K. pneumoniae *(I). Cyan arrows indicate KK 114 penetration inside the bacterial cytoplasm, resulting in higher fluorescence emission. A different magnification was used for each bacterial species. See Ref. [21] for labeling and imaging conditions.*

SYTO 9 and KK 114 emissions swiftly diminishes in response to colistin concentrations and treatment duration, whereas it remains constant in colistin-resistant strains [55]. This alternative method for detecting colistin resistance in *A. baumannii* offers a comparable duration (*i.e.*, less than 2 hours) to other techniques that are not based on bacterial cultivation and have a direct application in the clinical routine (*e.g.*, resazurin reduction-based assay [61].

The dye KK 114 served as an example of an intriguing case study, highlighting the importance of expanding the palette of available dyes in the development of new diagnostic and therapeutic approaches for treating bacterial infections. Encouraged by these promising results, we tested an extended set of dyes [21]. To elucidate the labeling mechanism, the dyes were used both in their reactive form (NHS esters) and inactive. Also importantly, the net charges varied in the range from +2 to −4. It was found that all the rhodamine dyes of the set and an exemplary oxazine dye KK 1905 (also as NHS ester), as well, can be used to label dead bacterial cells, thus expanding the range of available dyes for LIVE/DEAD applications. As we established, these dyes do not interfere with bacterial replication and can be used in live-cell imaging experiments, which is essential in bacteriology for gauging the physiological state of bacteria, untangling the intricate mechanisms at the basis of the bacterial divisome, performing real-time transcriptional analysis at the single-cell level, investigating phenotypic heterogeneity in bacterial populations, and exploring the host-bacteria interactions [62]. Among the tested dyes, only a few could effectively stain bacterial membranes. Out of the whole dye set, only KK 1517, KK 1518, and KK 1905-NHS could stain viable bacteria. In particular, the zwitter-ionic rhodamine KK 1517 ($z = 0$) was able to decorate the cell periphery of all the *B. subtilis* cells and of some *E. coli* cells, partially penetrating cytoplasm of both species. On the other hand, KK 1518 and KK 1905-NHS (both positively charged, see Ref. [21] for structures) selectively stained the cell periphery of *B. subtilis* and *E. coli* cells, with the latter generating brighter and more contrasted images. Note that the dye KK 1518 was a purposefully synthesized label with a net charge of +2 and a water-solubilizing moiety.

4. Conclusions

Very reasonable strategies were employed to develop user-friendly red and far-red emitting microscopy labels with tailored properties. Particularly, the authors summarize and discuss the syntheses of well-performing STED fluorophores with enhanced photostability. Some issues are highlighted that did not receive attention in the previous publications. Straightforward modifications at proper sites of a rhodamine dye scaffold enable synthesis and evaluation of diverse candidates. It was shown that the properties of the fluorophores alone, *i.e.*, polar functional groups and the fluorescence quantum yields, are not as crucial for STED microscopy as they were considered to be. As one can see from the comparative studies [11], the "race" for high quantum yields of fluorescence is not always justified. The photostability, proper instrument settings, the quality of antibodies, and the quality of labeling are the real deciding factors. Also, in recent years, new techniques were discovered that allow to reduce the high intensity of a STED beam and further increase the resolution, on the other hand, as one sees, for example, in [63, 64] and in the references therein. Importantly, apart from the conventional immunolabeling with antibodies or with HaloTag® ligands, there exists a great potential for non-covalent binding to cell structures, particularly bacterial membranes. The rhodamine dyes that were originally designed for STED microscopy a decade ago, for example, KK 114, have thus gotten a "new life" [21, 54]. This offers fresh avenues and rich toolboxes for microbiological research and diagnostic applications. Especially important could be custom-made fluorophores with multiple positive net charges [21], which were previously underestimated as STED dyes. Also, possibilities may be broadened in single-molecule studies and in medical diagnostics, as well. By diversifying the palette of the dyes available, researchers can enhance their ability to observe and analyze biological samples with greater precision and sensitivity, not necessarily using STED. This advancement holds promise for deeper insights into microbiology and related fields, as well as improving diagnostic and analytical techniques in medical and scientific research.

Acknowledgements

The authors are grateful to Prof. S. W. Hell for the opportunity to work at the Department of NanoBiophotonics (MPI-BPC Göttingen) and perform STED imaging in 2008—2015. Acknowledged are the grants, particularly, FKZ 13 N11066 and FKZ 13 N14122 (Bundesministerium für Bildung und Forschung), which financed the research work at that period. The studies would not be possible without extra images and good advice kindly provided by Dr. Daniel C. Jans, Dr. Franziska R. Winter, Dr. Fabian Göttfert, Dr. Maksim V. Sednev (MPI-BPC Göttingen) and Dr. Sergey M. Borisov (Graz University of Technology). Highly acknowledged are the European Society for Photobiology, the European Photochemistry Association, and the Royal Society of Chemistry for the license agreement allowing the authors to reproduce their figures. Very helpful were the good advice, instrumental, informational, and moral support from glyXera GmbH (Germany) and Peptide Technologies Ltd. (Russia). Dr. M. Lucidi especially acknowledges the support of NBFC to the University of Roma Tre (CN00000033).

Conflict of interests

The authors have no conflicts of interest to declare. None of the authors is employed by the companies whose products or instruments are mentioned.

Author details

Kirill Kolmakov[1]*, Massimiliano Lucidi[2,3] and Alexey V. Nizovtsev[4]

1 glyXera GmbH, Brenneckestraße, Magdeburg, Germany

2 Department of Science, Roma Tre University, Viale G. Marconi, Rome, Italy

3 NBFC, National Biodiversity Future Center, piazza Marina, Palermo, Italy

4 Gazpromneft – Industrial Innovations LCC, St. Petersburg, Russia

*Address all correspondence to: kirill_kolmakov@mail.ru

References

[1] Hell SW. Multicolor super-resolution imaging with photo-switchable fluorescent probes. Science. 2007;**317**:1749-1753. DOI: 10.1126/science.1146598

[2] Hell SW. Microscopy and its focal switch. Nature Methods. 2009;**6**:24-32. DOI: 10.1038/nmeth.1291

[3] Hell SW, Wichmann J. Breaking the diffraction resolution limit by stimulated emission: Stimulated-emission-depletion fluorescence microscopy. Optics Letters. 1994;**19**:780-782. DOI: 10.1364/ol.19.000780

[4] Boyarskiy VP, Belov VN, Medda R, Hein B, Bossi M, Hell SW. Photostable, amino reactive and water-soluble fluorescent labels based on sulfonated rhodamine with a rigidized xanthene fragment. Chemistry - A European Journal. 2008;**14**:1784-1792. DOI: 10.1002/chem.200701058

[5] Rust M J, Bates M, Zhuang X. Sub-diffraction-limit imaging by stochastic optical reconstruction microscopy (STORM). Nature Methods. 2006;**3**:793-795. DOI: 10.1038/nmeth929

[6] Huang B, Wang W, Bates M, Zhuang X. Three-dimensional super-resolution imaging by stochastic optical reconstruction microscopy. Science. 2008;**319**:810-813. DOI: 10.1126/science.1153529

[7] Huang B, Jones SA, Brandenburg B, Zhuang X. Whole-cell 3D STORM reveals interactions between cellular structures with nanometer-scale resolution. Nature Methods. 2008;**5**:1047-1052. DOI: 10.1038/nmeth.1274

[8] Fölling J, Bossi M, Bock H, et al. Fluorescence nanoscopy by ground-state depletion and single-molecule return. Nature Methods. 2008;**5**:943-945. DOI: 10.1038/nmeth.1257

[9] Kolmakov K, Belov V, Bierwagen J, Ringemann C, Müller V, Eggeling C, et al. Red-emitting rhodamine dyes for fluorescence microscopy and nanoscopy. Chemistry – A European Journal. 2010;**16**:158-166. DOI: 10.1002/chem.200902309

[10] Grimm JB, Xie L, Casler JC, et al. A general method to improve fluorophores using deuterated auxochromes. JACS Au. 2021;**1**(5):690-696. DOI: 10.1021/jacsau.1c00006

[11] Kolmakov K, Winter FR, Sednev MV, Ghosh S, Borisov SM, Nizovtsev AV. Everlasting rhodamine dyes and true deciding factors in their STED microscopy performance. Photochemical & Photobiological Sciences. 2020;**19**:1677-1689. DOI: 10.1039/d0pp00304b

[12] Wurm CA, Kolmakov K, Göttfert F, et al. Novel red fluorophores with superior performance in STED microscopy. Optical Nanoscopy. 2012;**1**:1. DOI: 10.1186/2192-2853-1-7

[13] Bachman JL, Pavlich CI, Boley AJ, Marcotte EM, Anslyn EV. Synthesis of carboxy ATTO 647N using redox cycling for xanthone access. Organic Letters. 2020;**22**(2):381-385. DOI: 10.1021/acs.orglett.9b03981

[14] Mao F, Leung W-Y, Haugland RP. Sulfonated xanthene derivatives. US Pat. 6130101. Molecular Probes Inc. 1997

[15] Abberior GmbH. Available from: https://abberior.shop/abberior-STAR-RED [Accessed: Feb. 16, 2024]

[16] Schmidt R, Wurm CA, Punge A, Egner A, Jakobs S, Hell SW. Mitochondrial cristae revealed with focused light. Nano Letters. 2009;**9**:2508-2510. DOI: 10.1021/nl901398t

[17] Stoldt S, Wenzel D, Kehrein K, Riedel D, Ott M, Jakobs S. Spatial orchestration of mitochondrial translation and OXPHOS complex assembly. Nature Cell Biology. 2018;**20**:528-534. DOI: 10.1038/s41556-018-0090-7

[18] Kolmakov K, Hebisch E, Wolfram T, et al. Far-red emitting fluorescent dyes for optical nanoscopy: Fluorinated silicon–rhodamines (SiRF dyes) and phosphorylated oxazines. Chemistry : A European Journal. 2015;**21**:13344-13356. DOI: 10.1002/chem.201501394

[19] Mitronova GY, Polyakova S, Wurm CA, et al. Functionalization of the meso-phenyl ring of rhodamine dyes through S_NAr with sulfur nucleophiles: Synthesis, biophysical characterizations, and comprehensive NMR analysis. European Journal of Organic Chemistry. 2015:337-349. DOI: 10.1002/ejoc.201403269

[20] Kolmakov K, Wurm CA, Meineke DNH, et al. Polar red-emitting rhodamine dyes with reactive groups: Synthesis, photophysical properties, and two-color STED nanoscopy applications. Chemistry : A European Journal. 2014;**20**:146-157. DOI: 10.1002/chem.201303433

[21] Lucidi M, Capecchi G, Visaggio D, et al. Expanding the microbiologist toolbox via new far-red-emitting dyes suitable for bacterial imaging. Microbiology Spectrum. 2024;**12**:e0369023. DOI: 10.1128/spectrum.03690-23

[22] Kolmakov K, Wurm CA, Hennig R, et al. Red-emitting rhodamines with hydroxylated, sulfonated, and phosphorylated dye residues and their use in fluorescence nanoscopy. Chemistry : A European Journal. 2012;**18**:12986-12998. DOI: 10.1002/chem.201201168

[23] Savicheva EA, Mitronova GY, Thomas L, et al. Negatively charged yellow-emitting 1-aminopyrene dyes for reductive amination and fluorescence detection of glycans. Angewandte Chemie, International Edition. 2020;**59**:5505-5509. DOI: 10.1002/anie.201908063

[24] Hell SW, Belov VN, Bischoff M, et al. Advanced Methods for Automated High-Performance Identification of Carbohydrates and Carbohydrate Mixture Composition Patterns and Systems Therefore as Well as Methods for Calibration of Multi Wavelength Fluorescence Detection Systems Therefore, Based on New Fluorescent Dyes. Eur. Pat. EP3914912A1. Max-Planck-Gesellschaft; 2019

[25] Romieu A, Brossard D, Hamon M, et al. Postsynthetic derivatization of fluorophores with alpha-sulfo-beta-alanine dipeptide linker. Application to the preparation of water-soluble cyanine and rhodamine dyes. Bioconjugate Chemistry. 2008;**19**:279-289. DOI: 10.1021/bc7003268

[26] Chevalier A, Piao W, Hanaoka K, et al. Azobenzene-caged sulforhodamine dyes: A novel class of 'turn-on' reactive probes for hypoxic tumor cell imaging. Methods and Applications in Fluorescence. 2015;**3**:044004. DOI: 10.1088/2050-6120/3/4/044004

[27] Romieu A, Tavernier-Lohr D, Pellet-Rostaing S, et al. Water solubilization of xanthene dyes by post-synthetic sulfonation in organic media. Tetrahedron Letters. 2010;**51**:3304-3308

[28] Romieu A, Bruckdorfer T, Clavé G, et al. N-Fmoc-α-sulfo-β-alanine: A versatile building block for the water solubilisation of chromophores and fluorophores by solid-phase strategy. Organic & Biomolecular Chemistry. 2011;**9**:5337-5342. DOI: 10.1039/c1ob05730h

[29] Massif C, Dautrey S, Haefele A, et al. New insights into the water-solubilisation offluorophores by post-synthetic "click" and Sonogashira reactions. Organic & Biomolecular Chemistry. 2012;**10**:4330-4336. DOI: 10.1039/c2ob25428j

[30] Chevalier A, Renard P-Y, Romieu A. Straightforward access to water-soluble unsymmetrical sulfoxanthene dyes: Application to the preparation of far-red fluorescent dyes with large stokes' shifts. Chemistry : A European Journal. 2014;**20**:8330-8337. DOI: 10.1002/chem.201402306

[31] Smit JH, van der Velde JHM, Huang J, et al. On the impact of competing intra-andintermolecular triplet-state quenching onphotobleaching and photoswitching kinetics of organic fluorophores. Physical Chemistry Chemical Physics. 2019;**21**:3721-3733. DOI: 10.1039/c8cp05063e

[32] Van der Velde JHM, Smit JH, Hebisch E, et al. Self-healing dyes for super-resolution fluorescence microscopy. Journal of Physics D: Applied Physics. 2019;**52**:034001

[33] Van der Velde JHM, Oelerich J, Huang J, et al. A simple and versatile design concept for fluorophore derivatives with intramolecular photostabilization. Nature Communications. 2016;**7**:10144. DOI: 10.1038/ncomms10144

[34] Van der Velde JHM, Uusitalo JJ, Ugen L-J, et al. Intramolecular photostabilizationviatriplet-state quenching: Design principles to make organicfluorophores"self-healing". Faraday Discussions. 2015;**184**:221-235. DOI: 10.1039/c5fd00114e

[35] Van der Velde JHM, Oelerich J, Huang J, et al. The power of two: Covalent coupling of photostabilizers for fluorescence applications. Journal of Physical Chemistry Letters. 2014;**5**:3792-3798. DOI: 10.1021/jz501874f

[36] Zheng Q, Jockusch S, Zhou Z, et al. On the mechanisms of cyanine fluorophore photostabilization. Journal of Physical Chemistry Letters. 2012;**3**:2200-2203. DOI: 10.1021/jz300670p

[37] Zheng Q, Jockusch S, Rodríguez-Calero GG, et al. Intra-molecular triplet energy transfer is a general approach to improve organic fluorophore photostability. Photochemical & Photobiological Sciences. 2016;**15**:196-203. DOI: 10.1039/c5pp00400d

[38] Klán P, Šolomek T, Bochet CG, et al. Photoremovable protecting groups in chemistry and biology: Reaction mechanisms and efficacy. Chemical Reviews. 2013;**113**:119-191. DOI: 10.1021/cr300177k

[39] Clegg RM. Förster resonance energy transfer - FRET what is it, why do it, and how it's done. In: Gadella TWJ, editor. Laboratory Techniques in Biochemistry and Molecular Biology. Vol. 33. Amsterdam: Elsevier; 2009. pp. 1-57. ISBN: 9780080549583

[40] Winter F, Loidolt M, Westphal V, et al. Multicolour nanoscopy of fixed and living cells with a single STED beam and hyperspectral detection. Scientific Reports. 2017;**7**:46492. DOI: 10.1038/srep46492

[41] Masters BR. The development of fluorescence microscopy. In:

Encyclopedia of Life Sciences (ELS). Chichester: John Wiley & Sons, Ltd; 2010. DOI: 10.1002/9780470015902.a0022093

[42] Toseland CP. Fluorescent labeling and modification of proteins. Journal of Chemical Biology. 2013;**6**:85-95. DOI: 10.1007/s12154-013-0094-5

[43] Dedecker P, De Schryver FC, Hofkens J. Fluorescent proteins: Shine on, you crazy diamond. Journal of the American Chemical Society. 2013;**135**:2387-2402. DOI: 10.1021/ja309768d

[44] Zhang J, Campbell RE, Ting AY, Tsien RY. Creating new fluorescent probes for cell biology. Nature Reviews. Molecular Cell Biology. 2002;**3**:906-918. DOI: 10.1038/nrm976

[45] Fernández-Suárez M, Ting AY. Fluorescent probes for super-resolution imaging in living cells. Nature Reviews. Molecular Cell Biology. 2008;**9**:929-943. DOI: 10.1038/nrm2531

[46] Parisi M, Lucidi M, Visca P, Cincotti G. Super-resolution optical imaging of bacterial cells. IEEE Journal of Selected Topics in Quantum Electronics. 2023;**29**:1-13. DOI: 10.1109/JSTQE.2022.3228121

[47] Berney M, Hammes F, Bosshard F, Weilenmann H-U, Egli T. Assessment and interpretation of bacterial viability by using the LIVE/DEAD BacLight kit in combination with flow cytometry. Applied and Environmental Microbiology. 2007;**73**:3283-3290. DOI: 10.1128/AEM.02750-06

[48] Stocks SM. Mechanism and use of the commercially available viability stain, BacLight. Cytometry Part A. 2004;**61A**:189-195. DOI: 10.1002/cyto.a.20069

[49] Freire JM, Gaspar D, de la Torre BG, Veiga AS, Andreu D, Castanho MARB. Monitoring antibacterial permeabilization in real time using time-resolved flow cytometry. Biochimica et Biophysica Acta (BBA) – Biomembranes. 2015;**1848**:554-560. DOI: 10.1016/j.bbamem.2014.11.001

[50] Stiefel P, Schmidt-Emrich S, Maniura-Weber K, Ren Q. Critical aspects of using bacterial cell viability assays with the fluorophores SYTO9 and propidium iodide. BMC Microbiology. 2015;**15**:36. DOI: 10.1186/s12866-015-0376-x

[51] Robertson J, McGoverin C, Vanholsbeeck F, Swift S. Optimisation of the protocol for the LIVE/DEAD BacLightTM bacterial viability kit for rapid determination of bacterial load. Frontiers in Microbiology. 2019;**10**:801. DOI: 10.3389/fmicb.2019.00801

[52] Robertson J, McGoverin C, White JR, Vanholsbeeck F, Swift S. Rapid detection of *Escherichia coli* antibiotic susceptibility using live/dead spectrometry for lytic agents. Microorganisms. 2021;**9**:924. DOI: 10.3390/microorganisms9050924

[53] Ni Y, Wu J. Far-red and near infrared BODIPY dyes: Synthesis and applications for fluorescent pH probes and bio-imaging. Organic & Biomolecular Chemistry. 2014;**12**:3774-3791. DOI: 10.1039/c3ob42554a

[54] Lucidi M, Hristu R, Nichele L, et al. STED nanoscopy of KK114-stained pathogenic bacteria. Journal of Biophotonics. 2020;**13**:e202000097. DOI: 10.1002/jbio.202000097

[55] Lucidi M, Hristu R, Nichele L, et al. Characterization of Acinetobacter baumannii filamentous cells by re-scan confocal microscopy and complementary fluorometric approaches. IEEE

Journal of Selected Topics in Quantum Electronics. 2021;**27**:1-7. DOI: 10.1109/JSTQE.2020.3048476

[56] Lucidi M, Imperi F, Artuso I, et al. Phage-mediated colistin resistance in Acinetobacter baumannii. Drug Resist Update. 2024;**73**:101061. DOI: 10.1016/j.drup.2024.101061

[57] Tacconelli E, Carrara E, Savoldi A, et al. Discovery, research, and development of new antibiotics: The WHO priority list of antibiotic-resistant bacteria and tuberculosis. The Lancet Infectious Diseases. 2018;**18**:318-327. DOI: 10.1016/S1473-3099(17)30753-3

[58] Sass P, Brötz-Oesterhelt H. Bacterial cell division as a target for new antibiotics. Current Opinion in Microbiology. 2013;**16**:522-530. DOI: 10.1016/j.mib.2013.07.006

[59] Vicidomini G, Bianchini P, Diaspro A. STED super-resolved microscopy. Nature Methods. 2018;**15**:173-182. DOI: 10.1038/nmeth.4593

[60] Schermelleh L, Ferrand A, Huser T, et al. Super-resolution microscopy demystified. Nature Cell Biology. 2019;**21**:72-84. DOI: 10.1038/s41556-018-0251-8

[61] Lescat M, Poirel L, Tinguely C, Nordmann P, Resazurin A. Reduction-based assay for rapid detection of polymyxin resistance in Acinetobacter baumannii and pseudomonas aeruginosa. Journal of Clinical Microbiology. 2019;**57**:e01563-e01518. DOI: 10.1128/JCM.01563-18

[62] Cambré A, Aertsen A. Bacterial vivisection: How fluorescence-based imaging techniques shed a light on the inner workings of bacteria. Microbiology and Molecular Biology Reviews. 2020;**84**. DOI: 10.1128/MMBR.00008-20

[63] Weber M, Leutenegger M, Stoldt S, et al. MINSTED fluorescence localization and nanoscopy. Nature Photonics. 2021;**15**:361-366. DOI: 10.1038/s41566-021-00774-2

[64] Remmel M, Scheiderer L, Butkevich AN, et al. Accelerated MINFLUX nanoscopy, through spontaneously fast-blinking fluorophores. Small. 2023;**19**:2206026. DOI: 10.1002/smll.202206026

Chapter 8

Colors in Forensics: The Analysis and Visualization of Forensic Data and Evidence

Tommy Bergmann, Ronny Bodach, Laura Pistorius, Svenja Preuß, Paul Seidel and Dirk Labudde

Abstract

In the realm of forensic science, the analysis and visualization of data and traces rely heavily on the interpretation of colors. This chapter delves into the multifaceted role colors play in forensic investigations, exploring their significance in various domains, such as bloodstain analysis, fingerprint examination, image forensics, and the study of artificial light sources. From discerning the age of bloodstains to identifying manipulated digital images, color analysis emerges as a pivotal tool in unraveling crime events and establishing facts crucial for legal proceedings. Through a discussion of selected forensic methods, this chapter highlights the diverse applications of color analysis and emphasizes the need for standardized approaches to ensure the accuracy and reliability of forensic investigations. Looking ahead, the continuous advancement of technologies and methodologies in color analysis promises to enhance further the efficacy of forensic science in solving crimes.

Keywords: forensic science, color analysis, digital image forensics, bloodstain age estimation, artificial light sources

1. Introduction

Forensics represents an intriguing discipline encompassing methodologies aimed at reconstructing crime events to facilitate their resolution [1]. A pivotal aspect of these methodologies involves the scrutiny of color properties or alterations in evidence.

Light and the associated perception of color have always been integral components of forensics [2]. One of the oldest tools used at crime scenes was oblique lighting, which still plays a major role in trace evidence analysis today [3]. The range of color analysis in forensics now spans the entire electromagnetic spectrum of light [2]. Well-known examples include making latent biological trace materials visible using UV light, evaluating or identifying vein patterns in the near-infrared range, and analyzing pigments in car paints.

Within this book chapter, we embark on a comprehensive exploration of various domains within forensic science, shedding light on the indispensable role colors play therein.

We commence by scrutinizing blood traces, often characterized by their distinct blood color associated with their chemical composition. Such analysis can furnish valuable insights into the timing of the crime, thus aiding in the reconstruction of events (Section 2.1). Another significant area is dactyloscopy, where fingerprints can be subjected to electrochemical excitation to glean additional insights into their origins. Color analysis assumes a crucial role in fingerprint identification and classification (Section 2.2). Image forensics emerges as a contemporary field dealing with the scrutiny of digital images. Here, accurate interpretation of color spaces is imperative to align the appearance of captured objects with reality as faithfully as possible. This analysis is pivotal in detecting manipulations or forgeries and verifying image authenticity (Section 2.3). Lastly, we delve into artificial light sources, which are pivotal in forensic analysis. Signals captured from such sources can furnish insights into the timing of the crime and the authenticity of recordings. Color analysis of these light sources is instrumental in drawing reliable conclusions from the recorded data (Section 2.4).

The generation, capture, and interpretation of color impressions, employing both established and novel methodologies, play a pivotal role in forensics. Within this chapter, we aim to spotlight selected potential applications of color analysis and underscore its significance in forensic analysis.

2. The term of color in selected forensic methods

Forensics encompasses the scientific discipline dedicated to investigating evidence and traces in criminal or legal contexts to ascertain facts, establish guilt or innocence, and aid in the resolution of crimes. It encompasses various subfields, including criminalistics, forensic medicine, digital forensics, forensic psychology, and forensic pathology. These disciplines entail the examination of physical evidence at crime scenes, the analysis of DNA and other biological materials, scrutiny of digital evidence such as computer data, and psychological profiling of perpetrators. The digital transformation, which also changes the nature and targets of crimes, in no way implies that color analysis is losing its significance. Even in the digital realm, material properties can be approximated using color information; it even offers new possibilities for the analysis of light and color. For this reason, color analyses in both analog and digital spaces will be presented below. Chapter 2 will delve into four specific methods, elucidating the role of colors in the generation, recording, or interpretation of trace materials. We commence with the analysis of biological evidence and culminate with the scrutiny of digital traces.

2.1 Analysis of age-related color changes using the example of forensic bloodstain age estimation

Within forensics, the analysis of blood traces constitutes a pivotal component in crime resolution. It furnishes investigators with vital clues for perpetrator identification and supplies information crucial for reconstructing the crime scene. The hue of blood emerges as a prominent factor, particularly in blood age estimation, wherein it serves as a basis for informed approximations regarding the age of blood droplets. This segment elucidates the rationale behind color alterations and provides a succinct overview of contemporary methods employed in blood color analysis.

2.1.1 Properties of blood

Blood is often revered as the quintessential essence of life, symbolizing the enduring vitality inherent in the human body. It is readily obtainable, and variations in its constituent components serve as crucial indicators of an individual's health status or, in forensic contexts, the age of blood trails [4].

Functionally, blood supports and sustains various bodily processes by facilitating the transport of essential substances. Mechanically propelled through the blood vessels by the rhythmic contractions of the heart, it traverses from the heart via arteries to organs and then returns via veins. On average, the human vascular system contains approximately 70–80 ml of whole blood per kilogram of body weight. Comprising roughly 55% blood plasma and 45% blood cells, including platelets, red blood cells, and white blood cells, whole blood predominantly consists of red blood cells. Constituting around 96% of the blood's mass, red blood cells (erythrocytes) serve as the primary carriers of oxygen absorbed from the lungs to organs and facilitate the transport of carbon dioxide from organs back to the lungs, thus facilitating cellular respiration [4].

The red color of blood stems predominantly from the protein hemoglobin, which constitutes the majority of red blood cell content. Hemoglobin comprises four subunits, each bound to an iron II complex known as heme. Central to the heme complex is an iron atom capable of reversible binding to oxygen molecules.

In the circulatory system, hemoglobin exists either in its oxygen-enriched form, oxyhemoglobin, or is deoxygenated during its journey back to the lungs as deoxyhemoglobin. Exposure to atmospheric oxygen outside the body gradually oxidizes oxyhemoglobin to methemoglobin, irreversibly converting the central iron atom to its trivalent state and binding water in place of oxygen. This oxidation process darkens the overall color of the blood. Over time, the concentration of methemoglobin increases, ultimately resulting in complete hemoglobin oxidation. As decay progresses, typically after 2–3 weeks, amino acids such as histidine tightly bind to the central iron atom, leading to the formation of hemichrome. This denaturation of blood pigment is accompanied by a color change from brown red to dark red or black.

Upon exiting the body, a drop of blood undergoes a gradual transformation in composition and subsequent color, transitioning from light red to deep black. The pace of this metamorphosis is contingent upon external factors, such as surface properties, temperature, or humidity [5].

2.1.2 Spectroscopic analysis of color change

Spectroscopy is a biophysical measurement method extensively employed across various forensic domains. It furnishes insights into the dimensions, configuration, architecture, charge, molecular mass, functionality, and kinetics of scrutinized macromolecules. This technique harnesses specific characteristics of light, enabling deductions regarding the condition of the analyzed sample.

More specifically, spectroscopy encompasses a collection of physical methods wherein the interactions between electromagnetic radiation and surrounding matter are observed and quantified. These interactions are recorded in the form of spectra using instruments called spectrometers. Spectra represent the intensity distributions of radiation as a function of wavelength, aiding in the identification of the substance under examination (**Figure 1**) [6, 7].

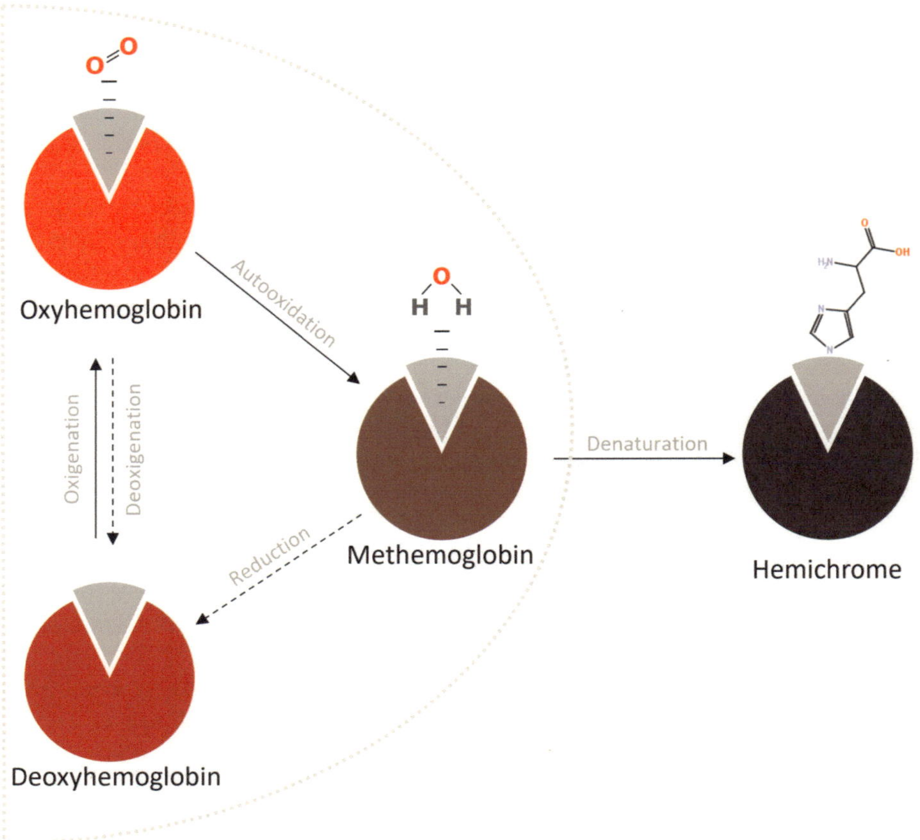

Figure 1.
Hemoglobin derivatives: the divalent iron present in oxyhemoglobin undergoes oxidation to trivalent iron, resulting in the formation of methemoglobin, which loses its ability to transport oxygen. Subsequently, during denaturation, histidine complexes frequently emerge, wherein the iron is fully bound, forming hemichromes. The distinct peaks exhibited by these hemoglobin derivatives in the electromagnetic spectrum contribute to the transition of blood color from light red to brown and ultimately to black (own illustration).

The aforementioned changes in blood, attributed to varying hemoglobin derivatives, can be monitored through blood spectra. These spectra exhibit evolving bands contingent upon the age and prevailing hemoglobin derivative. Notably, the Soret band, originating from the fundamental chemical structure of the heme group, exhibits a maximum intensity at approximately 425 nm in young bloodstains. As stains age, this peak progressively shifts toward the ultraviolet range, with blood traces older than 3 weeks showcasing a Soret peak around 400 nm. Additionally, young blood manifests peaks of oxyhemoglobin at 542 and 577 nm. However, as blood ages, the proportion of methemoglobin increases, characterized by peaks at 510 and 631.8 nm. Over time, oxyhemoglobin peaks diminish in intensity within the overall blood spectrum, giving way to methemoglobin peaks [8, 9].

In practice, forensic investigators collect blood samples from crime scenes and subject them to spectroscopic analysis. The spectral bands obtained are then mathematically compared with literature values to estimate the approximate age of the bloodstains. A detailed procedural elucidation of this methodology can be found in the chapter "Forensic Analysis of Bloodstain Color" from the book "Advances in Colorimetry" by IntechOpen (**Figure 2**).

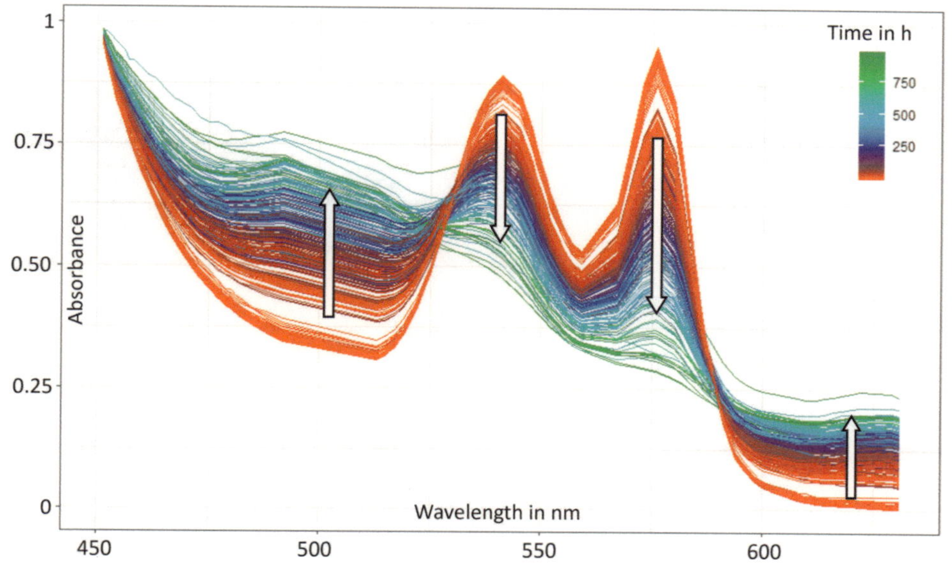

Figure 2.
Blood spectra depending on the stain age: the spectra of blood stains spanning various ages (ranging from 0.1 to 1000 hours) are compared, with color grading reflecting the age progression. age-associated alterations in the spectra are depicted at distinct positions, stemming from varying concentrations of hemoglobin derivatives (modified from [10]).

Ultimately, it is imperative to underscore that spectroscopic recording of the age-related chemical alteration in the central iron atom of the blood pigment hemoglobin is conducted to estimate the forensically relevant timeframe during which the blood traces have been present at the crime scene.

2.2 Analysis of electrochemically stimulated color changes using the example of a novel method for detecting fingerprints

This subchapter offers insights into a new research field within the realm of forensic dactyloscopy. Unique characteristics of the fingerprint were electrochemically stimulated, facilitating high-resolution capture of the trace and recording of forensically pertinent features.

Dactyloscopy—Characteristics of fingerprints Dactyloscopy, derived from ancient Greek (dáktylos "finger" and skopiá "peeping") [11], pertains to the examination of papillary ridges on the palms of hands and the soles of feet. The biometric method of identity verification through dactyloscopy, commonly known as fingerprinting, is based on the biological irregularities present in human papillary ridges.

Dactyloscopy has been utilized in forensic science for identification purposes since the early twentieth century, making it the oldest biometric method. Its introduction in Germany dates back to 1903 when Paul Koettig implemented it at the Dresden police headquarters. A notable criminal case in 1914 involved the murders on Hohlbeinstrasse and Terrassenufer in Dresden, where fingerprints left on a metal cassette were decisively linked to seamstress Marie Margarethe Müller.

Francis Galton presented various skin ridge patterns and devised a classification system, laying the groundwork for Edward Richard Henry's pattern classification system, which is still in use today. Fingerprinting captures an image of the papillary

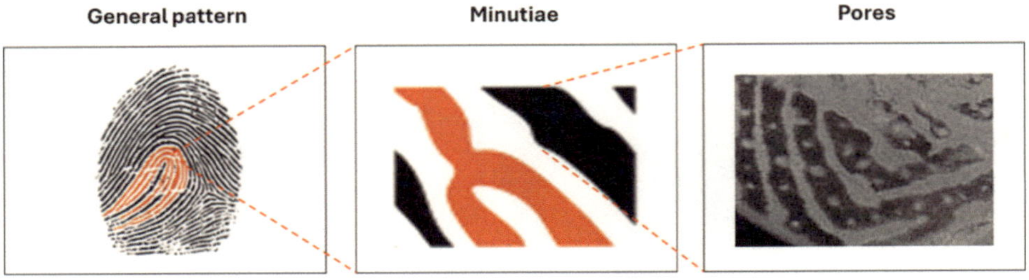

Figure 3.
Levels of information from a fingerprint that is used for identification (own illustration).

ridges on the fingertips, with each individual having a unique fingerprint. Biologically, a papillary ridge refers to an elevation in the epidermis on the palm or sole. Henry's classification system distinguishes various features of the fingerprint.

Figure 3 depicts one of the general patterns, anatomical features or "minutiae," and the sweat pore pattern, all of which can serve as identifying features in dactyloscopy.

Physical, chemical, or hybrid methods are employed to preserve fingerprints, often complemented by optical aids, such as oblique light, transmitted light, halogen, and UV light, as well as magnifying glasses [12].

Forensics essentially always involves the comparison of trace materials. In reality, this comparison is often hindered by the fact that the traces at the crime scene inherently have or due to the method of collection exhibit low quality. Therefore, methods are needed to preserve or improve this quality. Furthermore, it is always advantageous if traces can be examined directly at the crime scene, as an analysis following the prescribed "legal preservation" often leads to points of contention in court. Both of these issues are intended to be addressed with the novel electrochemical analysis method outlined below.

2.2.1 Electrochemistry in dactyloscopy

The electrochemical analysis method facilitates the generation of high-resolution images of papillary ridges, thereby enabling the examination of sweat pores. This significantly enhances the recording quality of fingerprints and the underlying anatomical features. Furthermore, it opens up the possibility of translating sweat pore assessment from fundamental research into practical forensic applications. With this innovative technology alone, basic patterns, minutiae, and pores are displayed in high resolution. The varying compositions of biomolecules on the fingerprint allow for gender determination and differentiation between adults and children. Additionally, insights into contact with chemical compounds such as explosives, drugs, or medication can be gleaned. Importantly, this method preserves the structural integrity of the fingerprint.

The novel analysis method relies on an electrochemical reaction. Initially, the fingerprint is mixed with non-destructive marker substances and then stimulated (luminescence). The application of voltage induces the release of electrons from a chemical compound to a reactant. The electronically excited marker substance subsequently returns to its baseline state by emitting light (luminescence). However, if grease or sweat is present on the fingerprint carrier, the reaction is impeded.

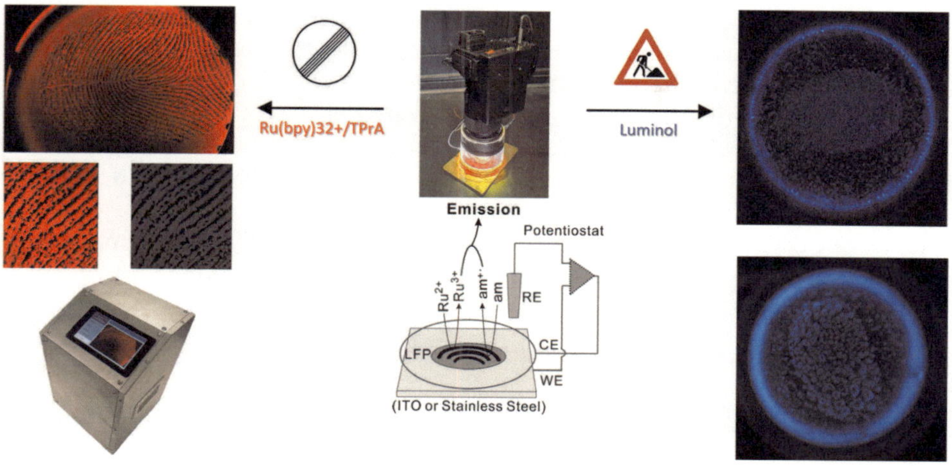

Figure 4.
From the idea to the prototype. The ECL imaging system, adapted from Xu et al., was utilized to electrochemically visualize fingerprints. While luminol enables this visualization, the formation of pronounced bubbles poses challenges for high-resolution trace evaluation. In contrast, analysis with the ruthenium complex proceeds smoothly, without such issues. Consequently, this variant was selected for creating the prototype depicted below (modified from [13]).

Consequently, only the spaces, that is, the finger grooves and pores, luminesce, create a highly detailed negative image of the fingerprint.

Alternatively, a positive image can be generated by treating the fingerprint with marker substances specific to the sought chemical compound (target substance). Subsequent application of the reaction partner results in luminescence only in areas where the desired compound is present. These may include narcotics, fire accelerants, or explosives. Relevant drugs within the scope of criminal law, such as cocaine, heroin (with the metabolite morphine or 6-monoacetylmorphine), codeine, methadone, and tetrahydrocannabinol (THC), can also be detected using this method.

For a comprehensive fingerprint analysis using this technique, the marker substance must target amino acids as the desired substance, causing the entire fingerprint to luminesce upon stimulation. Various substances can be employed for the basic electrochemical reaction. While luminol sometimes led to vigorous bubble formation on the trace, hindering analysis, ruthenium emerged as the ideal substance for the foundational electrochemical reaction.

These developed solutions culminated in the creation of an initial integrated solution comprising a modified camera and electrolytic cell, capable of implementing the desired functionalities. The high-contrast, high-resolution photos generated through this method could already be examined in the prototype using suitable image processing and evaluation programs (**Figure 4**) [14].

2.3 Image forensics

In recent years, there has been a significant shift toward digitalization worldwide. Nature, too, can now be digitally portrayed in myriad shapes and colors. In the field of image and video forensics, natural elements like people, landscapes, and objects are digitally captured, scrutinized, and analyzed. The initial stages of forensic investigation, marked by analog methods such as detecting bloodstains during aging or visualizing evidence at crime scenes, were just the beginning. The future lies in digital technologies, offering nearly boundless possibilities in terms of color variation for forensic examinations. The transition from analog to digital is often facilitated by cameras, which capture the world as we perceive it, thereby forming the basis for digital forensics, particularly in image and video analysis.

The capture of reality through digital devices also enables the recording of the entire color spectrum. This means that image and video analysis can provide a detailed description of individuals based on their color profiles.

However, cognitive comparison must be based on a forensic foundation, as the discretization of space can lead to deviations that need to be accounted for.

2.3.1 How a digital camera works/digital image collection

The design of a digital camera is modeled after that of the human eye, with various components of both systems carrying out similar functions.

Both the camera and the eye rely on external lenses to gather light, directing it onto the light-sensitive areas of their respective systems to achieve focus. Like the pupil and iris of the eye, the camera's aperture adjusts to control the amount of incoming light based on the surrounding lighting conditions. Finally, the process of converting light into electrical signals by the light sensors in a digital camera mirrors the function of cone and rod cells in the retina of the eye. These cells transform incoming light into electrical impulses, which are then further processed [15, 16].

2.3.2 Image capture process

In the realm of photography, what the human brain intuitively controls in the eye must be replicated through intricate engineering processes. The journey of information from a subject in an image through the sensor system to a digital format involves several complex steps (see **Figure 5**).

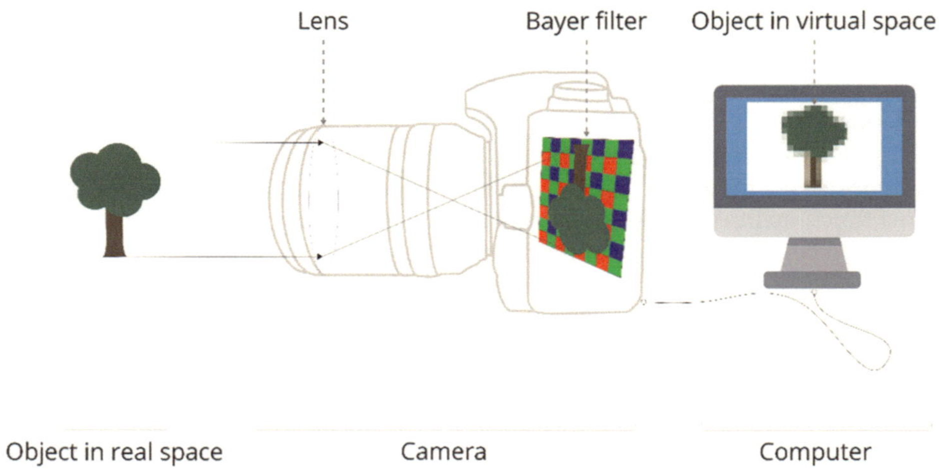

Figure 5.
From the object in real space to the object in virtual space. The light reflected by the real-space object is collected by the camera lenses, projected onto the image sensor through a Bayer filter and thereby converted into digital signals, which are subsequently saved as a digital image (own illustration).

DOI: http://dx.doi.org/10.5772/intechopen.1006108

Initially, the light reflected from the image subject is gathered by the camera's outer lenses. These lenses are pivotal in concentrating incoming light and projecting a sharp image onto the image sensor.

To achieve clarity in the image, the camera adjusts the distance between these outer lenses to ensure the subject is in focus. Typically, this adjustment is managed by a small motor, guided by measurements from an integrated distance-measuring device. This computer-controlled autofocus mechanism significantly contributes to producing clear, detailed, and non-blurry images.

At this juncture, the paths of the camera and the eye diverge. While the eye can only focus on a small area of the image at a time, with the brain synthesizing the overall scene from this limited information, the camera must simultaneously focus on multiple pixels of the object [17].

The light collected by the outer lenses reaches the camera's light sensor array, comprising millions of tiny photodiodes (pixels) that convert incoming light into electrical signals. To enhance image quality, each pixel is individually targeted by a corresponding microlens, with the number of microlenses aligning with the pixel count on the sensor. For instance, the iPhone 15 boasts a 48-megapixel camera, indicating a sensor array equipped with over 40 million lenses alone.

The functioning of a pixel resembles that of a photovoltaic cell, measuring the intensity of light it receives. This analog measurement (brightness) is then converted into a digital signal (brightness value). These sensors respond to light in the wavelength range of 190–1100 nanometers. This range encompasses ultraviolet (UV) and infrared (IR) light. Sensor sensitivity varies with material; while silicon sensors, common in most cameras, primarily respond to visible light, they can also react to other wavelengths, potentially causing visual noise if filtering is inadequate [18].

Since the light sensor detects the entire visible spectrum, filtering is necessary to capture not just brightness but also specific color values. This is achieved through millions of color filters distributed across the pixels, commonly organized according to the Bayer filter pattern [19]. The Bayer filter allows each pixel to capture only one primary color (red, green, or blue), enabling recording of both color and grayscale information. The Bayer filter arrangement is depicted in **Figure 5**.

Interpolation, also known as demosaicing or debayering, determines the color values of individual pixels on the image sensor. This process fills in missing color values through calculations involving neighboring pixels, resulting in each pixel having its own red, green, and blue values regardless of its original detection color.

However, interpolation may introduce artifacts such as blurred edges or color mixing, particularly in patterned subjects like a picket fence [20].

The outcome of this process is a RAW image, highly inefficient in terms of storage. Compression is often employed to mitigate storage requirements, albeit at the risk of data loss and image quality degradation.

2.3.3 Compression

To address the inefficiency of storing RAW images, compression algorithms like JPEG are commonly employed. The abbreviation "JPEG" stands for "Joint Photographic Experts Group," the organization behind this standard. The JPEG compression algorithm reduces image file sizes by eliminating redundant information and compressing image data without perceptible loss in quality to the human eye. This process involves techniques such as discrete cosine transformation (DCT) and quantization [21].

However, compression involves loss of information ("lossy compression"), particularly in areas with intricate detail or low contrast. This can lead to visible artifacts like blocking or blurring, especially in heavily or multicompressed images. Such artifacts are scrutinized in digital forensic analyses, particularly in detecting image forgery [22].

AI-assisted methods, especially in restoring images compressed multiple times with JPEG, add complexity to digital forensic analysis, a practice known as antiforensics [23].

2.3.4 Color representation in digital space

A digital image consists of a finite number of discrete elements called pixels. Each pixel is identified by a specific, predefined position within the image and holds one or more finite, discrete values [24].

Mathematically, a digital image (S) is commonly represented by a matrix $S = (s(x, y))$. The rows (L) of this matrix are known as image rows, and the columns (R) are referred to as image columns. The individual elements within this matrix correspond to pixels. The variable $x \in \{0, 1, 2, \ldots, L-1\}$ denotes the index of the respective row, while the variable $y \in \{0, 1, 2, \ldots, R-1\}$ denotes the index of the corresponding column indicates. Thus, s(x,y) represents the value of a pixel at the associated location coordinates [25].

2.3.5 Digital color spaces

In the simplest color space, each pixel contains only one intensity value, resulting in grayscale images. These images are represented as a two-dimensional matrix $S = (s(x, y))$, where each matrix element is assigned a value from a set of gray values G, typically $G = \{0, 1, 2, \ldots, 255\}$. Alternatively, values can be assigned to an interval from 0 to 1, where 0 represents black and 1 represents white by default [25].

In the context of color images with multiple channels, each pixel is represented not by a single value but by several color channels (N). Thus, s(x, y) becomes an N-dimensional vector. Multichannel images are represented by a three-dimensional image matrix S(s(x, y, n)), where the new variable specifies the index of the respective image channel [25].

The RGB color space (see **Figure 6**, left), one of the most well-known, is depicted as a unit cube in a three-dimensional Cartesian coordinate system. It is based on the primary colors red, green, and blue. The grayscale values range along the main diagonal from (0, 0, 0) for black to (255, 255, 255) for white [24].

This representation maintains the additive color mixing property, where white can be achieved through the maximum intensity of mixing red, green, and blue. Any other color can be represented by a combination of these primary colors. The mathematical representation, assuming each primary color represents a channel (n), would be s(s(x, y, 3)) with s(x, y) = (gRed, gGreen, gBlue) (modified from [25]).

Another prominent color model is the HSV color model (see **Figure 6**, right), where H represents hue, S represents saturation, and V represents intensity, each ranging from 0 to 1. This model is often visualized as an inverted hexagonal cone, where (gH, 0, 0) represents black and (gH, 255, 0) represents white [24].

An advantage of the HSV color space is the ability to analyze and compare hue independently of brightness and saturation, which is beneficial for forensic analysis.

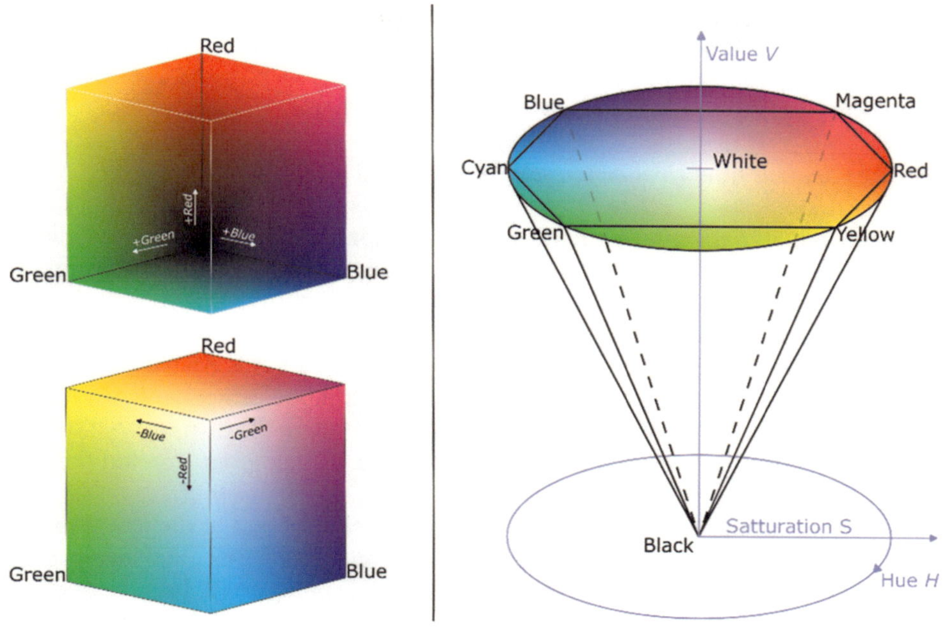

Figure 6.
RGB and HSV color model. Left: the RGB color space is depicted as a cube. In the top left, there is an RGB cube showing black in the bottom back corner and demonstrating the addition of RGB values. In the bottom left, another RGB cube is shown with white in the top, front corner, illustrating the subtraction of RGB values. Right: the HSV color space is represented as an inverted cone with six edges. The middle arrow indicates the intensity value, increasing in the direction of the arrow. Saturation is denoted by the lower arrow pointing to the right, with saturation increasing in the direction of the arrow. The color is described by a circle (own illustration).

2.3.6 White balance

Color shifts within the pixel color values of an image can result in the image appearing tinted, often dominated by a particular color. This effect is particularly noticeable in night shots, where the orange light from street lamps influences the captured image, causing a shift toward an orange color range. To address this issue, the white balance method is employed. While the human eye can adapt to changes in lighting and resulting color shifts automatically, cameras often struggle with this adjustment. Professional photographers often use color charts with known color values to precisely correct color shifts. Without such reference charts, achieving accurate white balance can be challenging and may only be approximated.

One method used for white balance adjustment is the "Gray-world algorithm." This method assumes that the average pixel value in an image is neutral gray (with a pixel color value of 128 in the RGB color space, represented as (128, 128, 128)) due to the normal distribution of colors. By analyzing the average color of the image, this algorithm estimates and corrects the color shift in the appropriate direction [26].

2.3.7 From the color to the case

Digital colors indeed hold significant importance in our daily lives, extending beyond the realm of multimedia. They also play a crucial role in casework, particularly in the development of specific evidence during criminal proceedings. For instance, colors can be utilized forensically to support various facts by comparing different hues to a desired color. Moreover, the ability to "make colors visible" is not merely fiction in certain cases.

2.3.8 Video analysis from rival motorcycle clans

The video material from Eisenbahnstrasse in Leipzig in 2017 depicts a volatile situation where two groups, identified by their distinctive black jackets, converge. The groups move toward each other, leading to a confrontation involving physical contact and gunfire. One individual is seen falling to the ground and remaining motionless. These groups are identified as the United Tribunes and the Hells Angels.

The footage, captured by a passer-by using a smartphone camera, suffers from poor quality and unstable camera work.

The initial question raised by the video material is: Which group entered the frame from the right, and which from the left?

The various groups are depicted in **Figure 7**, where a section of the image displays the logos on their jackets, albeit unclearly. Upon examining the original logos, it is difficult to definitively answer the question. However, in **Figure 8**, both logos are displayed clearly and scaled down to match the resolution of the image.

Simply comparing the shapes of the pixelated logos with the image sections taken may not provide sufficient information, but analyzing the colors can offer more insights. The standard Hells Angels logo incorporates the colors yellow and red. However, in the cellphone camera footage of the incident, various shades of gray are

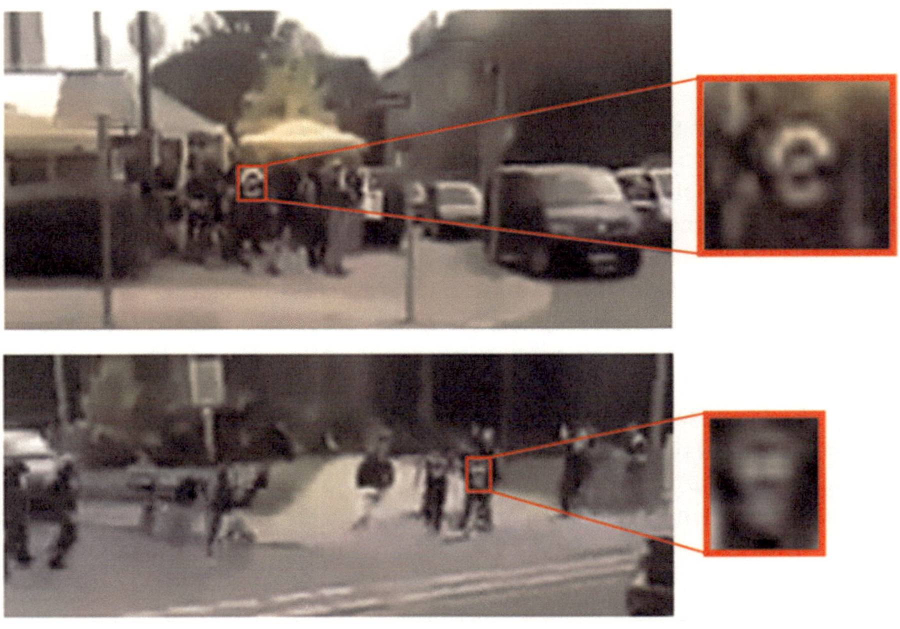

Figure 7.
Image sections from the video material, left-sided and right-sided grouping. The top image shows the left-sided group with an enlarged section of the group logo on a member's jacket. The bottom image shows the second group, also with an enlarged section of the corresponding group logo. The quality of the image sections is poor, making it impossible to clearly assign the persons to the groups represented (own illustration).

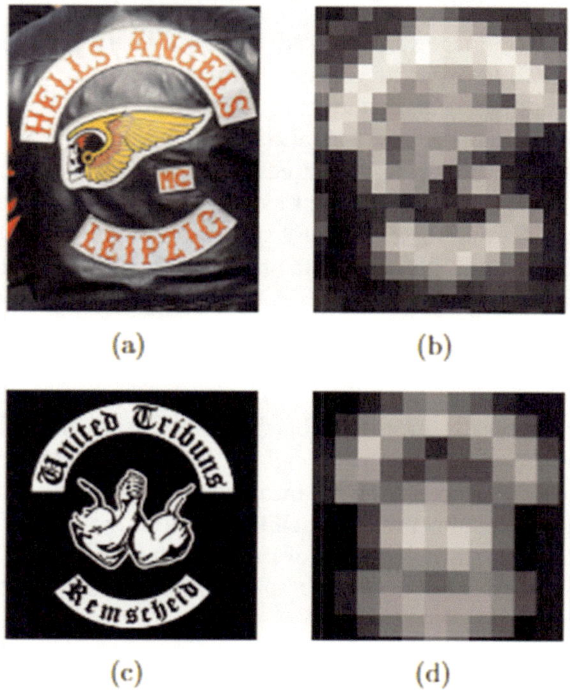

Figure 8.
Logos of the "Hells Angels" and "United Tribuns" groups. In comparison, here are the respective logos of the groups as they can be found on jackets. Image (a) shows the Hells Angels logo, which is scaled down to the quality of the video material to produce image (b). Image (c) corresponds to the logo of the United Tribuns. If this is scaled down, the result is image (d). The qualitative scaling is done to make the logos comparable with the footage from the video (own illustration).

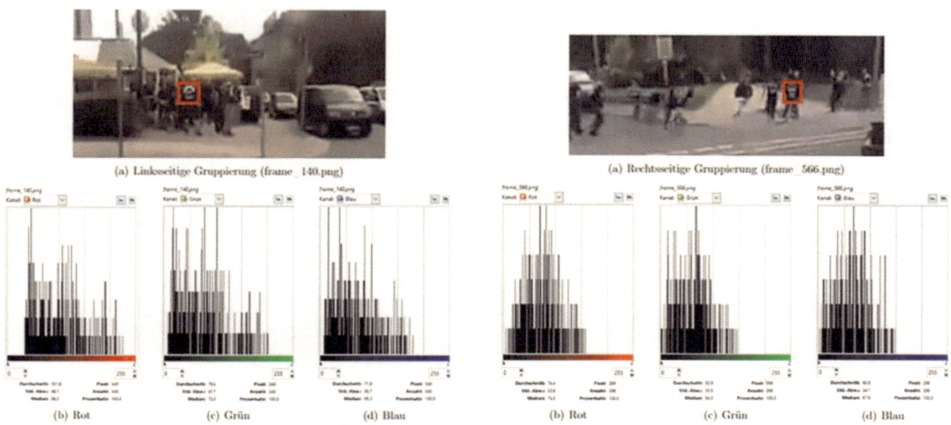

Figure 9.
Color spectrum analysis of the jacket logos from the video material. The application of the color spectrum analysis provides information about the color distribution of the corresponding image section. The representation of the individual color channels is listed here. The logo of the left-hand group produces the individual spectra on the left-, the right-hand diagrams correlate with the logo section of the right group (own illustration).

predominantly visible. Due to the low quality and resolution, the colors of individual areas are highly correlated, resulting in many pixels appearing gray. Nevertheless, the color spectra can illustrate the distribution and appearance of colors.

In **Figure 9**, the color spectra of both group logos are displayed. To achieve this, a section of the respective image was selected, focusing solely on the logo without including the surroundings or other objects in the analysis.

The color spectrum of the logo belonging to the group on the left exhibits a broad distribution across the colors red, green, and blue, ranging from less intense colors (appearing as black) to highly saturated areas. In contrast, the spectrum of individual colors in the logo of the group on the right extends less strongly into the intense color range. The differing characteristics of these spectra indicate that the logo on the left contains more color variation than the one on the right. Consequently, the spectrum on the left could be attributed to the Hells Angels group, as their logo typically incorporates colors as part of its standard design. Conversely, the logo of the group on the right displays much less intense colors, with the distribution mainly concentrated in the darker, black-appearing areas. Therefore, it could be associated with the United Tribuns logo, which typically appears in black and white without additional color.

Despite the limited information provided by the poor-quality video material regarding the colors of the logos, insights into color distribution can be gained through a color spectrum analysis. This method facilitated the assignment of the groups to their respective clans with a high degree of probability. It underscores the notion that just because colors may not be perceptible to the human eye in an image does not mean they are absent.

2.3.9 Identification of cars by color comparison from video material

In a scenario where a car drives on a road at night under low street lighting, surveillance cameras positioned across the street record the scene. Can digital forensics ascertain the color of the car and identify the make and model based on its paint job, especially when multiple models are available?

Specialized methods are employed to address this challenge, beginning with a color analysis of the car in the surveillance video. To accomplish this, one of the camera's

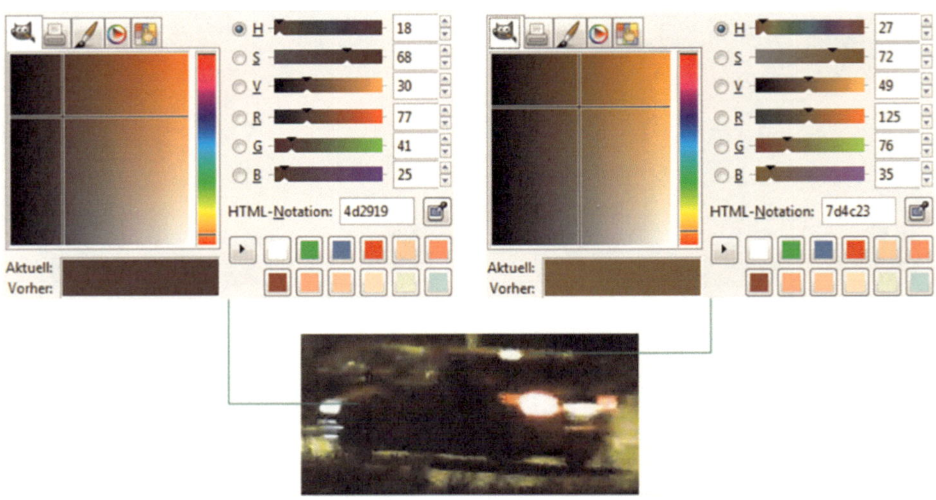

Figure 10.
The analysis reveals detail of a moving car exhibiting a visually perceived red color. Color information was extracted for a total of two pixels positioned around the image section. In both pixels, the red component is dominant over the green and blue components, suggesting a red car (own illustration).

images is loaded into an image processing program to extract color values. The analysis step is depicted in **Figure 10**.

The figure illustrates that the red component of the image surpasses the green and blue components. Color determination is not solely reliant on visual inspection but is also documented using analytical methods. However, there are various shades of red within the area of car painting. The subsequent step involves comparing the target vehicle with other cars of similar colors to ascertain its specific shade. Based on previous process steps, the focus has been on the Fabia vehicle type from the Skoda brand. **Figure 11** displays comparison models in various paint finishes within the red spectrum.

To generate meaningful comparison material, vehicles in the potential paint colors were driven past the original camera at night. The resulting video footage provides a more realistic basis for comparison. Upon comparing the underlying image or video material from the night of the crime with the comparison material, several differences emerge, including in the white balance. White values are closely linked to color temperature and brightness, factors that influence how objects are perceived by the human eye. For instance, cars may appear differently colored at various times of the day due to factors such as the position of the sun and atmospheric conditions like the refraction of sunlight by water molecules. The human brain naturally adjusts to interpret objects, such as a white sheet of paper or the pages of a book, as white under different lighting conditions.

However, a digital camera requires explicit instructions regarding the color temperature it should use, a process known as white balance. This adjustment was applied to the comparison video material.

Following adjustments to the video materials, the datasets of all comparison vehicles were prepared for comparison. Color matching of all vehicles with the original video material was conducted using a specialized image processing algorithm and quantitative assessment of color assignment. The results of these analyses, including correlations and differences, were summarized using an algorithm, with **Figure 12** presenting the results graphically.

The diagram illustrates a notably closer match between the Corrida Red paint color and the vehicle involved in the crime compared to the other types. Through the utilization of specialized color comparison algorithms and the subsequent analysis connecting them, it can be inferred with a high probability that among all the compared paint colors, Corrida Red is the one most likely present on the crime vehicle.

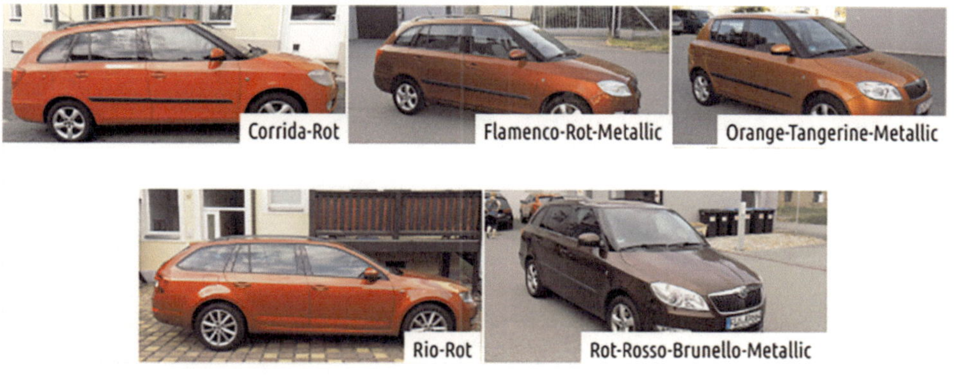

Figure 11.
Comparison models—depiction of several car types from the manufacturer Skoda, showcasing the comparison colors Corrida Red, Flamenco Red Metallic, Orange Tangerine Metallic, Rio Red, and Red Rosso Brunello Metallic (own illustration).

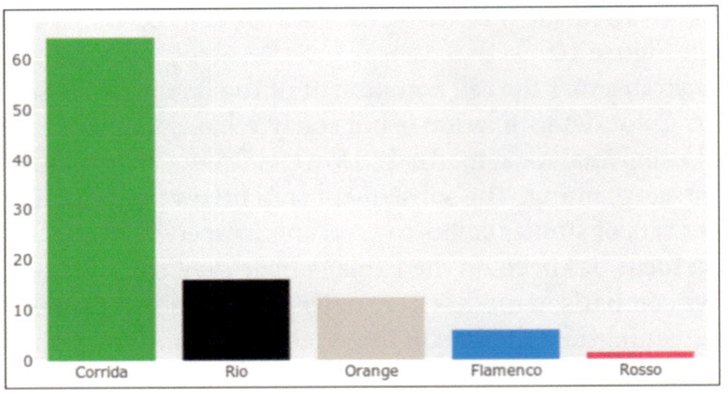

Figure 12.
The summary of all analysis data presents a comprehensive overview of correlations and differences between the comparison model images and the crime vehicle image. The shape of the bars in the visualization indicates the similarity of each comparison model image to the image of the crime vehicle. This allows for a clear understanding of the degree of resemblance and helps in identifying potential matches (own illustration).

2.3.10 The gold coin and the shoe

On March 27, 2017, the infamous Remmo clan stole the 100 kg gold coin "Big Maple Leaf" from the Bode Museum in Berlin. As part of the planning of this audacious theft, various individuals were captured on surveillance footage at Hackescher Markt a few days before the crime. Among them was a person wearing highly distinctive shoes, which bore a striking resemblance to a pair of shoes identified by the police during their investigation (see **Figure 13**).

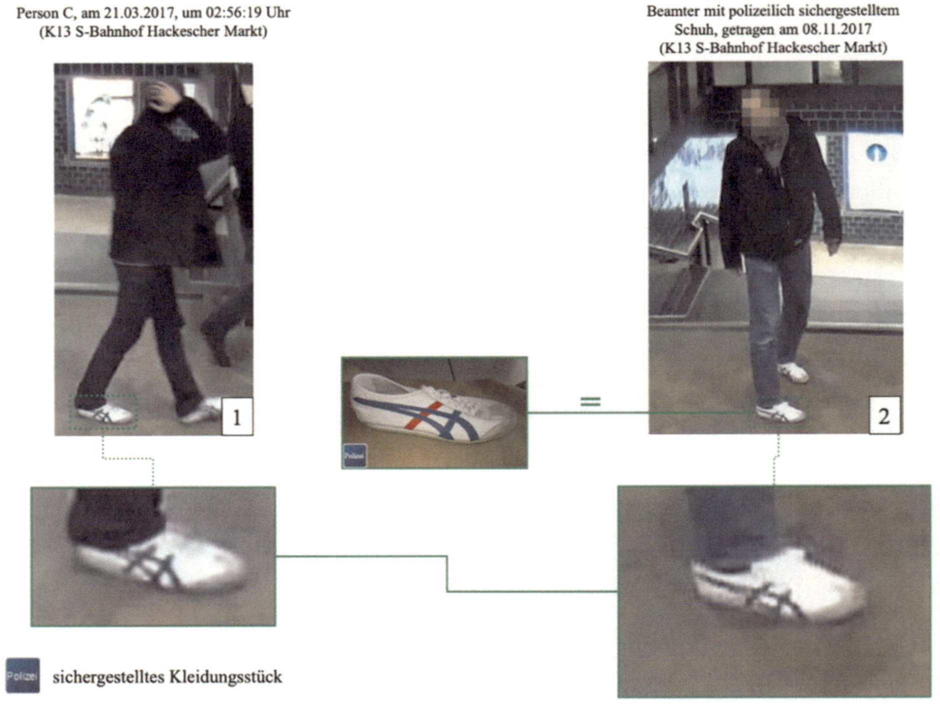

Figure 13.
(1) Suspicious person with distinctive shoes; (2) LKA officer with seized shoes (own illustration).

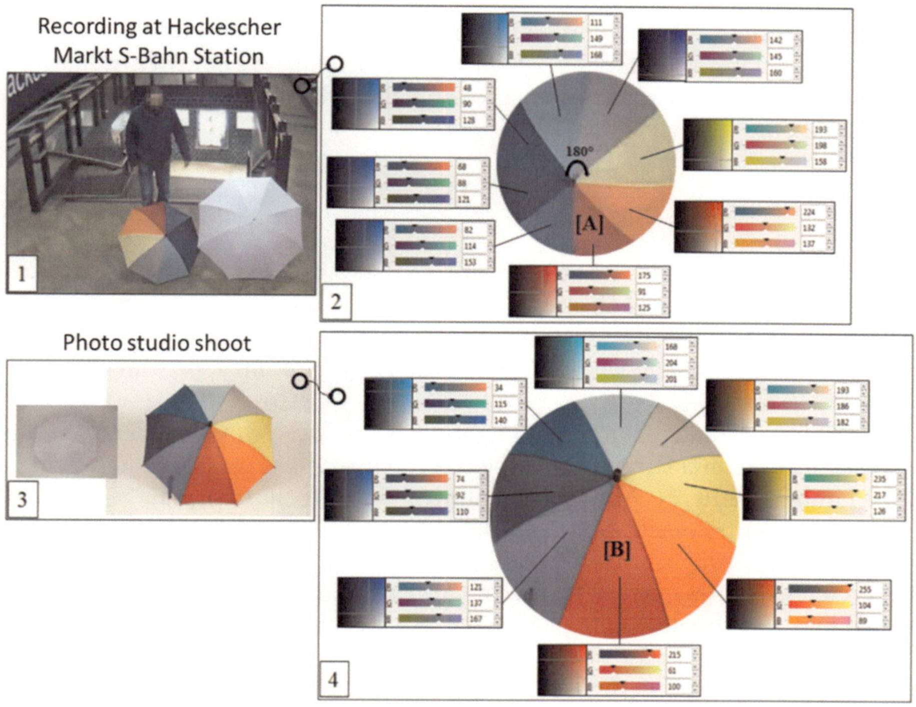

Figure 14.
(1, 2) Umbrellas at Hackerscher Markt; (3, 4) Umbrellas in the photo studio (own illustration).

The pattern of the comparison shoe (**Figure 13**, middle) precisely matches the pattern on the suspect's shoes (**Figure 13**, bottom left). However, a red stripe is clearly visible on the comparison shoe, which is not apparent in the video material. To investigate whether they are the same pair of shoes despite this discrepancy, comparative photos of the shoes worn by an LKA officer were captured using the same surveillance camera at the same location. This examination revealed that the surveillance camera loses some color information during recording and that, under these conditions, the comparison shoes at least appear similar to the suspect's shoes.

To quantify the color changes caused by the surveillance camera, two umbrellas were set up at Hackescher Markt and filmed: one multi-colored and one white. The same umbrellas were also photographed under studio conditions in a photo laboratory. These images were then compared for color and brightness values (see **Figure 14**).

This investigation unveiled that the surveillance camera inadequately captures red tones in recordings (see **Figure 14**, segments 2 and 4).

In this instance, it becomes evident how different camera models perceive colors differently. Accuracy in coloring is often not the primary concern with surveillance cameras, as it is often compromised in favor of considerations such as storage space, frame rates, or resolution. This fact frequently comes into play, particularly in correspondence analyses.

2.4 Artificial light sources as a source for detecting further forensic aspects

The preceding chapters have focused on the illumination of light sources and the utilization of artificial light and color sources. However, light itself can contain additional information that holds forensic value, particularly when artificial light sources are recorded. These recorded light sources can offer insights into the authenticity or chronology of the recordings. The examination of artificial light sources for forensic interpretation relies on signal processing methods used to analyze light fluctuations.

2.4.1 Creation of artificial light

Light comprises electric and magnetic fields that propagate in waves, constituting electromagnetic radiation or electromagnetic waves. Typically represented as a sine wave, light is characterized by wavelength, frequency, amplitude, and phase. The human eye perceives colors within the wavelength range from approximately 380 nm (violet) to 780 nm (red) [27].

Artificially generated light, termed lighting, is produced through energy conversion. Specifications of a lamp include the color temperature, measured in Kelvin (K), and the luminous efficacy, expressed in lumens per watt (lm/W) [28].

Fluorescent lamps operate on the principle of gas discharge and fluorescence technology. The energy supplied ionizes mercury atoms within the lamp, causing collisions with electrons, which emit ultraviolet radiation. This radiation is absorbed by a fluorescent coating inside the lamp, converting it into visible fluorescent light [29].

Due to phase changes in the AC network, the light intensity of artificial lighting fluctuates between maximum intensity and an off state. This fluctuation manifests as flickering at twice the main frequency, typically around 100 Hz in the European energy network [30].

2.4.2 European interconnected network

The public supply of electrical energy in Europe is facilitated by an interconnected system, comprising spatially neighboring and electrically connected transmission networks. Coordination and monitoring of this interconnected system are overseen by the association ENTSO-E (European Network of Transmission System Operators for Electricity) and its member transmission system operators [31]. As of March 2022, 36 European countries were integrated into this network system.

Since electrical energy cannot be stored sufficiently, energy consumption must align with production. Any imbalance between energy production and consumption leads to frequency deviations from the nominal frequency (fNenn) of 50.0 Hz in the power grid. The primary cause of these deviations is block-by-block electricity trading, although demand fluctuations also influence grid frequency. These deviations are constrained within a range of $\pm$200 mHz (millihertz), corresponding to a normal operating range of 49.80–50.20 Hz. The frequency and amplitude of these deviations vary depending on the time of day or region. For instance, strong positive frequency deviations are typically observed before 5 a.m., while significant drops in network frequency often occur in the evening.

Moreover, due to the 15 or 30-minute trading intervals in the European electricity grid, particularly pronounced fluctuations are expected between these intervals. There is a recurring pattern in the network frequency, characterized by a sharp drop at the start of a 15 or 30-minute interval, followed by an increase until the interval's conclusion [32].

DOI: http://dx.doi.org/10.5772/intechopen.1006108

2.4.3 Network frequency analysis

Network fluctuations in the nominal frequency can be assessed using Electric Network Frequency Analysis (ENF). One of the pioneering works on ENF, dating back to 2005, is by Grigoras [33], which addresses network frequency artifacts in audio files recognized as forensically significant. The ENF criterion primarily involves measuring the mains frequency at a power source to establish a reference data set, extracting the ENF from audio material, and subsequently comparing both ENF tracks.

Initially, the research primarily focused on applications in the field of audio forensics. Grigoras' approach outlines the extraction of an ENF track from audio recordings for timestamp verification. This involves downsampling to 120 Hz and band-limiting the audio signal to a frequency range of 49–51 Hz. The resulting ENF track is then visually compared with a reference database using Fast Fourier Transform (FFT) techniques [33].

Establishing a reference database for future analysis is crucial for classifying recorded ENF tracks. This entails selecting diverse measurement locations to capture frequency curves and reference data from various network areas.

2.4.4 Artificial light sources in videos

Video recordings from surveillance cameras, smartphones, digital cameras, and TV recordings are predominantly audiovisual, capturing both video images and accompanying audio tracks simultaneously. Additionally, digital video recordings are characterized by several parameters. These include:

1. Frame rate: the frame rate, measured in Hertz (Hz) or frames per second (fps), indicates the number of video frames recorded per unit of time.

2. Image size: this refers to the dimensions of the video frame, typically specified in terms of vertical pixels (columns) and horizontal pixels (rows), which determine the resolution of the video.

3. Color depth: the color depth specifies the number of color channels in an image, indicating the maximum number of possible color and brightness values per pixel.

The combination of these characteristics defines the video format of a recording. It is important to distinguish the video format from the data format used, which determines the technical structure and interpretability of the digital video data during processing. Typically, the data structure is organized as a container format containing separate audio and video content. This structure describes the compression methods or algorithms, known as codecs, used for encoding and decoding the audio and video content (**Figure 15**).

The global electronic shutter mechanism used here is called a global shutter. Compared to complementary metal-oxide-semiconductor (CMOS) sensors, charge-coupled device (CCD) sensors are more sensitive to light, have lower noise, and have a lower image readout rate. With CMOS sensors, pixels are exposed line by line from top to bottom and converted directly into a voltage signal and read out by an electronic circuit for each pixel element [31].

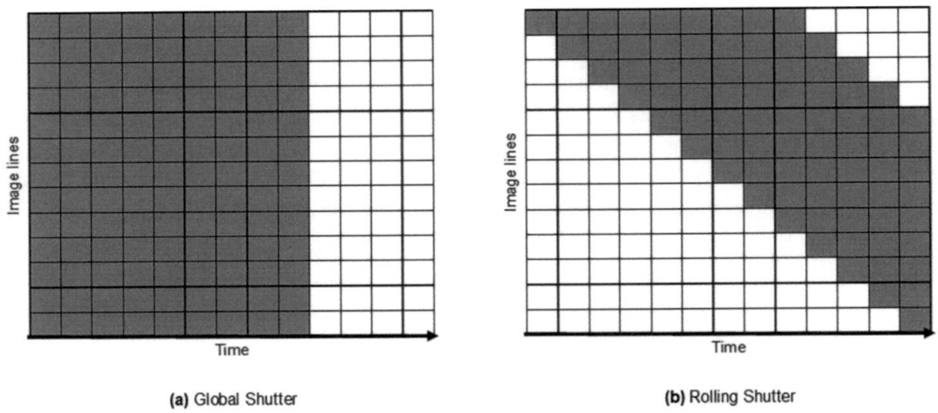

Figure 15.
How (a) CCD sensors and (b) CMOS sensors work (own illustration based on [34]).

The "rolling" shutter mechanism used, known as a rolling shutter, increases the image readout rate, but the line-by-line scanning also represents a source of error, for example when recording fast-moving objects. If the entire pixel area is scanned line or column by line slower than the movement of the recorded object, a distorted image is created due to the time delay in scanning from the first to the last line. This error is called the rolling shutter effect [35].

2.4.5 ENF in imagery and "white wall" method

Through the analysis of audio components in video recordings, an Electric Network Frequency (ENF) analysis can also be conducted for video files. In cases where the audio track is absent or unsuitable for ENF analysis, the video content captured under artificial lighting can serve as an alternative. Garg *et al.* presented a method in [29] 2011 to detect ENF using optical sensors and video cameras. This approach broadened the scope of ENF forensic investigations to encompass the examination of image material within video files. Studies by Garg *et al.* in [29] revealed that fluctuations in light intensity, caused by emissions from fluorescent lamps commonly used in indoor lighting, are directly correlated with ENF.

Garg *et al.* introduced the "white wall" method, wherein a white wall serves as the constant backdrop of a video and is illuminated by a fluorescent lamp. The average intensity of video frames over a 10-minute period was calculated, and a bandpass filter with a frequency band of ± 0.5 Hz around the calculated "aliased" frequencies fa was applied. Subsequent bandpass filtering revealed the ENF signal in spectrogram displays. Spectrogram analysis was conducted using the short-time Fourier transform, with parameters set to a window size of 480 video frames (approximately 16 seconds) and an overlap factor of 50%, facilitating the detection of ENF signals [29].

The results reported by Garg *et al.* underscore the necessity of calculating aliased frequencies f_a due to discrepancies between the fluorescent lamp's light flicker frequency ($f_l \approx 100$ Hz) and the camera's sampling rate (e.g., f_s = 30 fs = 30 Hz). This mismatch violates the sampling theorem, resulting in aliasing effects. To translate recorded video intensities from the time domain to the frequency domain, Fourier transformation is employed. However, since Fourier transformation can only detect frequencies between 0 and the Nyquist rate (f_{Nq}), aliased frequencies f_a must be determined for the light flicker frequency of 100 Hz [36].

The formula below, using the nominal light flicker frequency of 100 Hz and the camera's sampling rate is employed to calculate aliased frequencies [37]:

$$f_a = |f_l - k \cdot f_s| \leq \frac{f_s}{2}, \exists k \in \mathbb{N} \quad (1)$$

The aliased frequencies indicate the frequency ranges within which the ENF track can be localized [38].

The occurrence of the rolling shutter effect, associated with CMOS camera sensors, further affects the accuracy of ENF analysis detection. A problem arises due to time delays affecting the sequential exposure and reading of image material [39]. Experiments with various cameras revealed that an idle period follows the sequential line-by-line reading of video images before the next video image is read out [40]. Consequently, the ENF signal is not continuously captured. It was also observed that sequential image capture can contribute to an increase in the effective sampling rate, as multiple points in time are captured within a single video image [39]. This directly impacts the camera sampling rate of CMOS-based devices.

Adjustments to the sampling frequency are necessary to account for this phenomenon [41]:

$$f_{s_{cmos}} = F_{1 \times h} \cdot f_s \quad (2)$$

Examining the formula, it becomes evident that the product of the height $F_{1 \times h}$ (in pixels) of a video image and the video frame rate f_s (in fps) yields an increased sampling rate fscmos for videos recorded with CMOS sensors [41].

2.4.6 Comparison of network frequencies with databases

A comparison of the data between the recorded reference data and the data determined from the audio or video content can be performed by calculating the correlation coefficients. The correlation coefficient is a measure used in signal processing to quantify the similarity between two spatial or temporal functions, X and Y. Depending on whether the correlation coefficient is closer to +1 or −1, it indicates a positive or negative correlation between the data series. A positive correlation suggests that the measured values in the comparison file and the original file increase or decrease simultaneously. Conversely, if the correlation coefficient is closer to 0, it indicates little or no correlation.

To compute the correlation, the two signals under consideration, the one from the reference database and the ENF signal recorded from the video, are compared. To determine the time of recording in the reference data set, the shorter signal, the network frequency in the video recording, is systematically shifted relative to the longer signal, the network frequency of the reference database. The peak of the correlation function then indicates the maximum similarity between both signals.

Establishing the temporal reference of the network frequency in the reference recording enables the classification and determination of the recording time of the video, akin to a forensic timestamp of a video file.

The forensic utility of ENF analysis can be categorized into the following areas of application:

1. Determination of recording times through temporal analysis of reference data.

2. Identification of the recording locations by correlating different network connection areas.

3. Assessment of continuity/authenticity by identifying the presence of ENF signals.

2.4.7 Conclusion

With the increasing prevalence of video recordings generated using artificial intelligence or manipulated multimedia data, such as deepfakes, ENF analysis emerges as a crucial tool for verifying their authenticity [42]. Detecting deepfakes and other manipulations of audiovisual media files is essential in the fight against terrorism [43]. Additionally, these techniques are invaluable for solving various criminal acts related to video material. Examples include crimes involving ransom demands, kidnapping scenarios, attacks on surveillance systems, and the widespread distribution of child pornography video files [43].

3. Discussion and outlook

The discussed methods underscore the importance of color analysis and color itself in the reconstruction of criminal events, as well as the identification of suspects and crime-relevant objects. Color analysis is applied in various contexts, such as estimating the age of bloodstains and linking suspects to crime scenes. Similarly, suspect identification is aided by quantifying color changes in surveillance cameras and analyzing color spectra in logos. The classification of cars based on colors and their comparison in videos is also a significant application area.

Furthermore, the significance of colors in case processing in criminal proceedings is highlighted, with various facts being forensically supported through color matching. It is evident that different camera models perceive colors differently, emphasizing the need to consider this in forensic analysis.

An essential aspect to consider is the need for a standardized approach to color analysis in forensic investigations. Given the relatively new methods for analyzing digital color phenomena, there is often a lack of a gold standard or comparison values for satisfactory evaluation. Large-scale studies on data collection and analysis could enhance accuracy and comparability in the future. Additionally, the development of guidelines and best practices for color analysis in forensics could improve the quality and reliability of investigations.

It has been demonstrated that light itself can contain additional information with forensic value, especially when artificial light sources are recorded.

In summary, this work illustrates the diverse applications of color analysis in forensics and demonstrates how colors can support various forensic methods in drawing meaningful conclusions from recorded data. Continued advancements in technologies and methods for color analysis, coupled with the establishment of standards and best practices, will play a vital role in further advancing forensic science and aiding in crime-solving efforts.

4. Conclusion

Color and forensics are two inseparable fields. The information generated from color-based investigations and analyses is sometimes significant for various areas of analog and digital casework. Electromagnetic light allows insights into the real course of events in many ways and supports the task of truthfully reconstructing events. In particular, its applicability in the digital realm, for example during video analysis, supports our legal system on a new level. The digitalization of our everyday lives will continue to advance, and forensics will follow suit. The medium of color will also play an indispensable role in its relevance for digital forensic analyses.

Conflict of interest

The authors declare no conflict of interest.

Author details

Tommy Bergmann*, Ronny Bodach, Laura Pistorius, Svenja Preuß, Paul Seidel and Dirk Labudde
Forensic Science Investigation Lab (FoSIL), University of Applied Sciences Mittweida, Germany

*Address all correspondence to: bergmann@hs-mittweida.de

References

[1] Roux C, Crispino F, Ribaux O. From forensics to forensic science. Current Issues in Criminal Justice. 2012;**24**(1): 7-24. DOI: 10.1080/10345329.2012.12035941

[2] Miolo G, Stair J, Zloh M. Light in Forensic Science: Issues and Applications. Burlington House, Piccadilly, London: The Royal Society of Chemistry; 2018. DOI: 10.1039/9781788010344

[3] Lennard C, Stoilovic M. Application of Forensic Light Sources at the Crime Scene in "the Practice of Crime Scene Investigation". Boca Raton, Florida: CRC Press; 2004, eBook ISBN: 9780429245718

[4] Schmidt RF, Thews G. Physiologie des Menschen. Berlin, Heidelberg: Springer; 1997. S.411-435/604-606, ISBN 978-3-662-09347-4

[5] Bremmer RH et al. Biphasic oxidation of oxy-hemoglobin in bloodstains. PLOS One. 2011;**6**(7):e21845

[6] Schmidt W. Optische Spektroskopie: Eine Einführung für Naturwissenschaftler und Techniker. Weinheim: VCH Verlagsgesellschaft mbH; 1994. ISBN: 9783527290352

[7] Böcker J. Spektroskopie - Kapitel 1 (Einführung S. 26-61), 1. Aufl. Vogel Buchverlag; 1997. ISBN 3-8023-1581-2

[8] Patterson D. Use of reflectance measurements in assessing the colour changes of ageing bloodstains. Nature. 1960;**187**:688-689

[9] Zijlstra W, Buursma A, Assendelft VO. Visible and Near Infrared Absorption Spectra of Human and Animal Haemoglobin. VSP BV: Zeist; 2000

[10] Bergmann T, Leberecht C, Labudde D. Analysis of the influence of EDTA-treated reference samples on forensic bloodstain age estimation. Forensic Science International. 2021;**325**:4-5. DOI: 10.1016/j.forsciint.2021.110876

[11] Gemoll W. Griechisch-Deutsches Schul- und Handwörterbuch. München/Wien: G. Freytag Verlag; 1965

[12] Hamilton. Fingerprints. In: Houck MM, editor. Forensic Fingerprints. Amsterdam u. a: Elsevier; 2016. pp. 19-26, ISBN 978-0-12-800573-6

[13] Xu L et al. Imaging latent fingerprints by electrochemiluminescence. Angewandte Chemie (International Ed. in English). 2012;**51**(32):8068-8072. DOI: 10.1002/anie.201203815

[14] Bergmann T et al. Development and empirical optimization of an electrochemical analysis cell for the visualization of latent fingerprints and their chemical adhesives. In: International Conference of the Biometrics Special Interest Group (BIOSIG). Darmstadt, Germany: IEEE; 2020

[15] Ferland M. Comparison of the Human Eye to a Camera, Sciencing. 2018. Available from: https://sciencing.com/comparison-human-eye-camera-6305474.html [Accessed: March 8, 2024]

[16] National Keratoconus Foundation. How Does the Human Eye Work? 2019. Available from: NKCF.org. https://nkcf.org/about-keratoconus/how-the-human-eye-works/

[17] Cambridge in Color. Cameras Vs. the Human Eye. n.d. Available from: https://www.cambridgeincolour.com/tutorials/cameras-vs-human-eye.htm

[18] Hero G. Introduction to Light Emitting Diode Technology and Applications. Boca Raton, Florida: CRC Press; 2008. p. 116. Ch. 5. ISBN 1-4200-7662-0

[19] Bayer B. Color Imaging Array. Patent US 3971065A. 1976

[20] Buades A, Coll B, Morel JM, Sbert C. Non Local Demosaicing. New York City: IEEE TIP; 2007

[21] Wallace GK. The JPEG still picture compression standard. IEEE Transactions on Consumer Electronics. 1992;**38**(1):xviii-xxxiv

[22] Bianchi T, Piva A. Image forgery localization via block-grained analysis of JPEG artifacts. IEEE Transactions on Information Forensics and Security. 2012;**7**(3):1003-1017

[23] Jiang J, Zhang K, Timofte R. Towards flexible blind JPEG artifacts removal. In: IEEE/CVF International Conference on Computer Vision. Montreal: IEEE; 2021

[24] Chanda B, Majumder DD. Digital Image Processing and Analysis. New Delhi: PHI Learning Private Limited; 2011

[25] Nischwitz A, Fischer M, Haberäcker P. Bildverarbeitung—Band II des Standardwerks Computergrafik und Bildverarbeitung. 4. Auflage. Wiesbaden: Springer Fachmedien Wiesbaden; 2020

[26] Buchsbaum G. A spatial processor model for object colour perception. Journal of the Franklin Institute. 1980; **310**(1):1-26

[27] Löffler-Mang M. Optische Sensorik, Lasertechnik, Experimente, Light Barriers (Studium Fertigung), ger, 1. Aufl. Wiesbaden: Vieweg + Teubner; 2011, 2012. 244 S. ISBN: 3834814490

[28] Hering E. Sensoren in Wissenschaft und Technik, Funktionsweise und Einsatzgebiete. Wiesbaden: Springer Vieweg. in Springer Fachmedien Wiesbaden GmbH; 2012. ISBN: 9783834886354

[29] Garg R, Varna AL, Wu M. "Seeing" ENF: Natural time stamp for digital video via optical sensing and signal processing. In: Proceedings of the 19th ACM International Conference on Multimedia—MM '11, Scottsdale, Arizona, USA. ISBN: 9781450306164. 2011. DOI: 10.1145/2072298.2072303

[30] Konstantin P. Praxisbuch Energiewirtschaft. Berlin, Heidelberg: Springer Berlin Heidelberg. 590 S. ISBN: 978-3-662-49822-4; 2017. DOI: 10.1007/978-3-662-49823-1

[31] Jaschinsky M. Untersuchung des Zusammenhangs zwischen gemessener Netzfrequenz und Regelenergieeinsatz als Basis eines Reglerentwurfs zum Intraday Lastmanagement. 2012. Available from: http://hdl.handle.net/20.500.12738/6120

[32] Schäfer B et al. Non-Gaussian power grid frequency fluctuations characterized by Lévy-stable laws and superstatistics. Nature Energy. 2018; **3**(2):119-126. PII: 58. DOI: 10.1038/s41560-017-0058-z

[33] Grigoras C. Digital audio recording analysis: The electric network frequency (ENF) criterion. International Journal of Speech, Language and the Law. 2005;**12**(1):63-76. ISSN: 1748-8885. DOI: 10.1558/sll.2005.12.1.63

[34] Le T, Le N-T, Jang YM. Performance of rolling shutter and global shutter camera in optical camera

communications. In: International Conference on Information and Communication Technology Convergence (ICTC). Jeju Island: IEEE; 2015

[35] Schmidt U. Professionelle Videotechnik. Berlin, Heidelberg: Springer. 926 S. ISBN: 978-3-662-63943-6; 2021. DOI: 10.1007/978-3-662-63944-3

[36] Baksteen T. The Electrical Network Frequency Criterion, Determining the Time and Location of Digital Recordings [Masterarbeit]. Delft, Netherlands: Delft University of Technology; 2015. 65 S

[37] Hajj-Ahmad A et al. Flicker forensics for pirate device identification. In: Proceedings of the 3rd ACM Workshop on Information Hiding and Multimedia Security. Portland, Oregon, USA: Association for Computing Machinery; ISBN: 9781450335874; 2015. DOI: 10.1145/2756601.2756612

[38] Vatansever S, Dirik AE, Memon N. Detecting the presence of ENF signal in digital videos: A superpixel-based approach. IEEE Signal Processing Letters. 2017;**24**(10):1463-1467. ISSN: 1070-9908. DOI: 10.1109/LSP.2017.2741440

[39] Su H et al. Exploiting rolling shutter for ENF signal extraction from video. In: 2014 IEEE International Conference on Image Processing (ICIP). Paris, France: IEEE; 2014. pp. 5367-5371. ISBN: 978-1-4799-5751-4. DOI: 10.1109/ICIP.2014.7026086

[40] Vatansever S, Dirik AE, Memon N. Analysis of rolling shutter effect on ENF based video forensics. IEEE Transactions on Information Forensics and Security. 2019;**14**(9):2262-2275. ISSN: 1556-6013. DOI: 10.1109/TIFS.2019.2895540

[41] Nagothu D et al. Authenticating video feeds using electric network frequency estimation at the edge. ICST Transactions on Security and Safety. 2021;**168**:648. DOI: 10.4108/eai.4-2-2021.168648

[42] Nagothu D et al. Deterring deepfake attacks with an electrical network frequency fingerprints approach. Future Internet. 2022;**14**(5):125. PII: fi14050125. DOI: 10.3390/fi14050125

[43] Garg R. Time and Location Forensics for Multimedia. Department of Electrical and Computer Engineering [Dissertation]. College Park, Maryland: University of Maryland; 2013. 212 S

Chapter 9

Synthesis of Dyes and the Textile Industry

Joshua Inanabor Isibor

Abstract

In a country like Nigeria, the textile and chemical industry have experienced a regrettable decline and deterioration in the last two decades. This should be a major concern to the industrialist, chemist, governments and all other stakeholders in the industry. This chapter seeks to explain some basic processes in the synthesis of some dyes (Azo Dyes) with different colours from a single aniline compound-2,4-dichlorophenyl aniline (diazo component) severally reacted with five different coupling agents, namely: Phenol, Salicylic acid, 1-Naphthol, Resorcinol and Aniline, in the appropriate acidic and basic media to synthesise five different azo dyes. Cellulosic fabrics were successfully dyed with the synthesised dyes. The dyes showed good light and wash fastness properties on the cellulosic fabrics, which showed they have good affinity for such materials. The percentage yield, melting point, chromatography and refractive index, as well as UV-Visible Spectroscopic analysis were rightly carried out and reported. The synthesised dyes were also recommended for commercialization.

Keywords: industry, azo dye, diazo component, coupling agent, light and wash fastness

1. Introduction

The dye 'Mauve' was serendipitously discovered by William Henry Perkin in 1856 [1]. This discovery (of Mauveine) in 1856 is said to have laid a foundation for the synthetic organic chemical and dyestuff industry, [2] and this was about the same time modern chemistry began to evolve. Other *aniline* dyes followed, such as *fuchsine*, *safranin*, and *induline*. Many thousands of synthetic dyes have since also been prepared. Thus, it suffice to say that dye chemistry as an aspect of organic chemistry, is one in which the earliest scientific theories were first being used. However, this early scientific innovative which revolutionised the world of fashion and colour [3] has now been largely relegated to the background in our world today [4]. In fact, many of our academic and teaching institutions pay little attention to this vitally important branch of organic chemistry [5]. Dyes form a foundation for our different colours which in their various expressions have made our world aesthetically appealing. Colours derive from the spectrum of light interacting in the eye with the spectral sensitivities of the light receptors [6] and can as well be defined and quantified by the degree to which these receptors or cells are being stimulated. It should be noted that with regards to their solubility, organic colourants fall into two classes, namely: dyes and pigments [7]. The major distinction between them is that while dyes are soluble in water and/or organic solvents, pigments are insoluble in both types of liquid media [7].

2. Dyes

Dyes are coloured substances capable of application in aqueous or non-aqueous solution, in an aqueous dispersion to a substrate, so that the substrate acquires a coloured appearance. The substrates are usually a textile fibre, but may also be paper, hair, fur, plastic material, wax, cosmetic base or a foodstuff—from newer discoveries. Worthy of note also is that unlike most organic compounds, dyes possess their characteristic colours because they: (1) absorb light in the visible spectrum (400–700 nm), (2) have at least one chromophore (colour-bearing group), (3) have a conjugated system, i.e., a structure with alternating double and single bonds, and (4) exhibit resonance of electrons which is a stabilising force in organic compounds [8].

Groups in dye molecules which give them their characteristic colours are called chromophores [9]. These chromophores enable the dyes to absorb light of certain wavelengths within the visible spectrum, and examples of such groups are the –N=N– (Azo group), –NO (Nitroso Group), $-NO_2$ (Nitro Group) and –C=O (keto Group). Generally, molecules which contain chromophores are referred to as chromogens. Electron donors (e.g. –NH2, –NHMe, =NMe) and electron acceptors (–SH, OH, COOH) called auxochromes, also help intensify or deepen dye colours and also increase their solubility, by shifting the wavelength of light to a longer one [10].

2.1 Classes of dyes

2.1.1 Natural dyes

These are dyes derived from plants, invertebrates, or minerals. The majority of natural dyes are vegetable dyes from plant sources like roots, leaves, and wood; as well as other organic sources such as fungi and lichens. The archaeologists have found evidence of textile dyeing dating back to the Neolithic period. In China, dyeing with plants, barks, and insects has been traced back to more than 5000 years. Natural dyes actually need chemicals called mordants, to bind them to the textile fibres [11] and those used by early dyers ranged from tannin from oak-galls, vinegar, and ammonia from stale urine. Many mordants, and some dyes themselves produced strong odours, and large-scale dye workers were often isolated in their own districts. Typical examples of natural dyes recorded to being used since historical times include the natural invertebrate dyes like the *Tyrian purple* and crimson *kermes*, as well as plant-based dyes such as *woad* (*Isatis tinctoria*), *saffron*, indigo and *madder* usually found around Africa, Europe and Asia. Others include the *cochineal* and *logwood* (*Haematoxylum campechianum*) dyes some of which were highly priced luxury items at those times. The discovery of man-made (synthetic) dyes in the mid-19th century triggered a long decline in the large-scale market for natural dyes, however, due to health and environmental concerns earlier in the 21st century, the market for natural dyes in the fashion industry is beginning to experience a resurgence [12].

2.1.2 Synthetic dyes

These are human-made dyes, the first of which was the 'Mauvine', produced in the 1800s by Williams Henry Perkin [13]. Synthetic dyes with the 'Mauve' as the first [14] quickly replaced the traditional natural dyes over time.

Man-made dyes actually cost less, offer a vast range of new colours and import better properties to the dye materials. Almost all the colours seen today in various designs on textile materials are synthetic dyes. The synthetic dyes have today evolved into a multi-billion dollar industry but only in a few climes of the world at present.

The synthetic dyes can be named according to the chemical structure of their particular chromophoric groups. For example, diphenylmethane compounds, Oxazine compounds, Xanthene compounds. Other examples according to solubility and chemical properties include the acid dyes such as the Alizarine Pure Blue B and Acid red 88; basic dyes used for acrylic fibre and paper dying; reactive dyes like Drimarene K and Procion MX; Vat dyes like the Indigos and Sulphur dyes like the Sulphur Black 1. Azo dyes are examples of the most popular varieties of synthetic dyes. Today, they are being used up to 90% in the dyeing units, as they offer a wide variety of colours simple to synthesise. In addition to this, a wide range of colours can be obtained through the pre-treatment process.

2.2 Azo dyes

Azo dyes are characterised by presence in the molecule of one or more of the azo group N=N–, which form bridges between organic residues, of which at least one is usually an aromatic nucleus [7]. The azo groups are normally attached to sp^2-hybridised carbon atoms such as benzene, naphthalene, thiazole, and thiophene. Under normal conditions, azo dyes exist in the more stable planar trans-form so that the carbon-nitrogen bond angle is approximately 120° [15].

It is worthy of note that azo dyes are by far the most important class, and account for more than 60–70% of the total commercialised disperse dyes in the world [16]. Examples include the Solvent Yellow 7, C.I. Disperse Red 1, Trypan Blue, and so on. The reasons azo dyes have dominated the markets and have replaced many of the conventional anthraquinone dyes are not far-fetched. In principle, the azo group is an inherently intense chromophore in terms of the tinctorial strength; when compared comparatively with weak anthraquinone chromogens, normally two (2) or three (3) times stronger in tinctorial strength [4]. In addition, simplicity and ease of manufacture, unlike the other types of dyestuffs such as anthraquinone and styryl make the cost of manufacturing of azo dyes relatively lower than the anthraquinone dyes (**Figure 1**) [15].

2.3 A typical azo dye

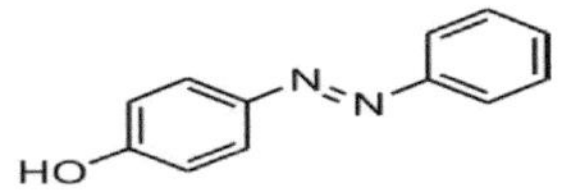

Figure 1.
Solvent yellow 7 [17].

3. Materials and method

50 ml breakers, 50 ml standard flask, buchner filter assembly, funnel, 50 ml conical flask, heater, mechanical stirrer, measuring cylinder, magnetic stirrer, gallenkamp melting point apparatus, capillary tubes, thermometer.

Chemicals and Reagents: sodium hydroxide, concentrated sulphuric acid, sodium nitrite, 2, 4-dichloroaniline, phenol, salicylic acid, 1-naphthol, resorcinol, aniline, ethanol.

3.1 Synthesis of 4- [2, 4-dichlorophenyl)azo]-phenol

2, 4-dichloroaniline (1.62 g) was dissolved in about 4.5 ml of 6 moles of acid (concentrated Sulphuric Acid). This solution was then poured onto cold water and cooled to 0°C. The coupling agent phenol (1.0 g) was dissolved in about 3.5 ml of 5 M Sodium Hydroxide. To the diazo-compound-acid-cold water mixture was then added 5 ml Of 5 M sodium nitrite for diazotization. The Phenol-Sodium- Hydroxide solution was added to the diazotizing mixture with vigorous stirring for about 1 hour.

The crude precipitate (7.33 g) was filtered, washed with water, and dried.

Re-crystallisation from boiling ethanol gave deep orange powder (6.12 g), 83.5%. Melting point (195°C).

TLC: indicated one spot.

Rf = 0.71 (Ethanol).

3.2 Synthesis of 4- [2, 4-dichlorophenyl)azo] – salicylic acid

2,4-dichloroaniline (1.62 g) was dissolved in concentrated Sulphuric Acid (4.7 ml). This mixture was mixed with cold water and the entire mixture cooled to 0°C. The coupling agent, Salicylic Acid (1.4 g) was dissolved in 5 M Sodium Hydroxide (3.5 ml). 5 M Sodium Nitrite (3.6 ml) was then added to the diazo compound-cold water mixture; after which the coupling agent mixture was added with vigorous stirring, for 1 hour.

The crude precipitate (7.64 g) was filtered, washed with water and dried.

Re-crystallisation from boiling ethanol gave deep orange dye (6.37 g), 83.38%.

Melting point (170°C).

TLC: indicated one spot.

Rf = 0.71 (Ethanol).

3.3 Synthesis of 4-[2,4-dichlorophenyl)azo]-1-naphthol

2, 4-Dichloroaniline (1.62 g) was dissolved in concentrated Sulphuric Acid (4.5 ml). This solution was then mixed with cold water and the mixture cooled to 0°C. The coupling agent 1-Naphthol (1.4 g) was then dissolved in 5 M Sodium Hydroxide (3.6 ml). 5 M Sodium Nitrite (3.6 ml) was then added to the diazo component mixture. The coupling agent mixture was then added, with vigorous stirring for 1 hour.

The crude precipitate (8.03 g) was filtered, washed with water, and dried.

Re-crystallisation from boiling ethanol gave deep orange powder (6.61 g), 82.3%.

Melting point (178°C).

TLC: indicated one spot.

Rf = 0.71 (Ethanol).

3.4 Synthesis of 4-[2, 4-dichlorophenyl)azo] - resorcinol

2, 4-Dichloroaniline (1.62 g) was dissolved in concentrated Sulphuric Acid (4.5 ml). This solution was then mixed with cold water and the mixture cooled to 0°C. The coupling agent- Resorcinol (1.0 g) was dissolved in 5 M Sodium Hydroxide (3.6 ml). 5 M Sodium Nitrite (3.6 ml) was then added to the diazo component mixture, after which the coupling agent mixture was added with vigorous stirring for 1 hour.

The crude precipitate (6.33 g) was filtered, washed with water and dried.

Re-crystallisation from boiling ethanol gave yellow powder (5.33 g), 82.62%.

Melting point (178°C).

TLC: indicated one spot.

Rf = 0.72 (Ethanol).

3.5 Synthesis of 4-[2, 4-dichlorophenyl)azo]-aniline

2,4-Dichloroaniline (1.62 g) was dissolved in concentrated Sulphuric Acid (4.5 ml). This solution was mixed with cold water and the mixture cooled to 0°C. The coupling agent – Aniline (1.0 g) was dissolved in 5 M Sodium Hydroxide (3.6 ml). 5 M Sodium Nitrite (3.6 ml) was added to the diazo component mixture, after which the coupling agent mixture was added with vigorous stirring for 1 hour.

The crude precipitate (7.34 g) was filtered, washed with water and dried.

Re-crystallisation from boiling ethanol gave pale red powder (6.28 g), 85.6% (**Table 1**).

Melting point (179°C).

TLC: indicated one spot.

Rf = 0.78.

The respective structures of the 2, 4-dichlorophenyl Aniline Series Synthesised are shown below:

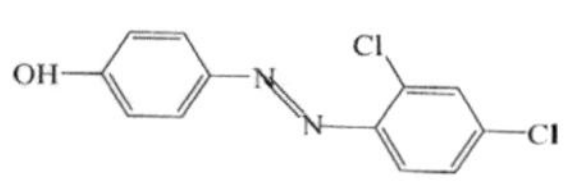

: 4- (2, 4-dichlorophenylazo)-Phenol.

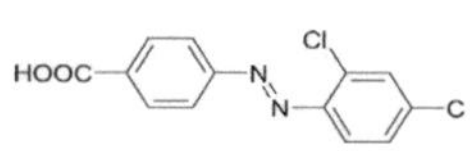

: 4- (2, 4-dichlorophenylazo)-salicylic acid.

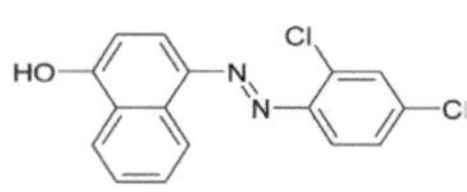

: 4- (2, 4-dichlorophenylazo) -1-naphthol.

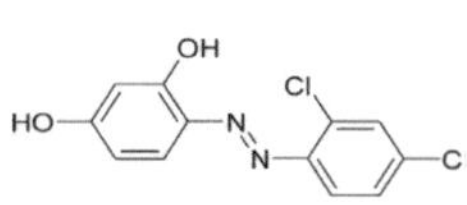

: 4- (2, 4-dichlorophenylazo) – resorcinol.

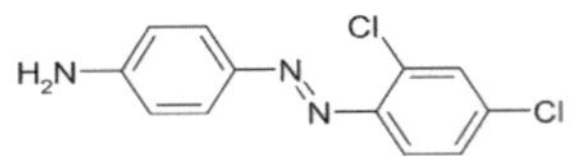

: 4- (2, 4-dichlorophenylazo) – aniline.

Dye	Wave length (λ_{max})	Absorbance (A)	Conc (Mol/L) (E_{max})	Molar Absorptivity
4-[2, 4-Dichlorophenyl) azo]-Phenol	675	2.430	1×10^{-4}	24,300
4- [2, 4-Dichlorophenyl) azo]-Salicylic Acid	20,200	540	2.020	1×10^{-4}
4- [2, 4-Dichlorophenyl) azo]-1-Naphthol	25,600	675	2.560	1×10^{-4}
4- (2, 4-Dichlorophenyl) azo]-Resorcinol	620	2.316	1×10^{-4}	23,160
4-[2,4-Dichlorophenyl) azo]-Aniline	19,720	540	1.972	1×10^{-4}

Table 1.
Characterisation result of the UV–visible spectroscopic analysis for the 2, 4-dichlorophenyl aniline dye derivatives.

4. Discussion

The azo dyes produced provided a range of colours from deep orange to milk, yellow and red. The dyes were generally observed to have high colour intensity, brightness and depth on dyeing on cellulosic fabric. The nature and position of substituents present on the diazotized compound and respective coupling agents had influence on the depth and shades of the synthesised dyes [18]. The auxochromic groups (e.g. -0H-, -COOH-) in the dye structures resulted in bathochromic shifts which led to the high intensity and depth of the produced dyes, with good light and wash fastness with cellulosic fibre. The fastness properties of dyes are correlated with their structure and substrate to which they are applied [19]. In addition, the melting point values of these 2, 4- dichloroaniline dye derivatives were observed to be generally high, ranging from 170 to 210°C, gave one spot on the chromatographic plate, have a high refractive index and percentage yield. This indicates a high level of purity which is evident in the brightness of the synthesised dyes. The nomenclature of produced dyes reveals that groups in the coupling component were generally para-directing, and ortho for resorcinol. The greatest effects in terms of longer wavelength is achieved by placing the substituent in the positions ortho or para to the azo group for effective conjugation [20]. This also explains for the large wavelengths the synthesised dyes (540–675 nm). This is within the wavelength in the visible spectrum in which azo dyes generally absorb light (400–700 nm) [9].

5. Conclusion

The variety of intensely deep and bright colours derivable with various coupling agents' reactions with a single diazo-component makes the synthetic dye like the azo(s) a very good option for optimal commercialisation by the dye and textile industries. The chemical and textile industry can be revived through conscious efforts by all major players in the industry. Government can expedite actions, enact laws and reinforce existing but dormant acts, engineered towards a revamp of the chemical and subsequently textile industries for meaningful contribution to the nation's gross domestic product (G.D.P), and general comfort of the citizenry.

Acknowledgements

I am firstly grateful to God Almighty for the privilege, inspiration and provisions in making this research work a reality from start to finish. My sincere appreciation especially goes to my selfless and honourable research supervisor, *Professor D. Ukponmwan* of the University of Benin, Benin-City, Edo State, Nigeria; who mentored and guided me through every step in this research work journey. He recommended me to the relevant research laboratories of the Chemistry Department, University of Benin to be provided with all available reagents and especially apparatus needed for the full success of my research work. I am further grateful to my parents, family and spouse for their unreserved provision and support through this research journey and writing. I earnestly pray that God immensely and abundantly rewards them all in entirety.

Author details

Joshua Inanabor Isibor
Federal College of Forest Resources Management, Fugar, Edo State, Nigeria

*Address all correspondence to: joshuainanabor2@gmail.com

References

[1] Berradi M et al. Textile finishing dyes and their impact on aquatic environs. Heliyon. 2019;**5**:e02711

[2] Sivagami M, Saharan IV. Phyto-mediated Ni/NiO NPs and their catalytic applications-a short review. Inorganic Chemistry Communications. 2022;**145**:110054

[3] The color purple: How an accidental discovery changed fashion forever. CNN. 2018. Available from: https://edition.cnn.com/style/article/perkin-mauve-purple/index.html

[4] Kassia SC. The Secret Lives of Colour. London: John Murray; 2016. p. 21. ISBN 9781473630819. OCLC 936144129

[5] Gruyter D. Organic Chemistry in Colour. Berlin, Boston: De Gruyter; 2022. ISBN: 9783112541746

[6] Berns Roy S. In: Billmeyer FW, Saltzman M, editors. Billmeyer and Saltzman's Principles of Color Technology. 4th ed. Hoboken, NJ: Wiley; 2019. pp. 5-9. ISBN 978-1119366683. OCLC 1080250734

[7] Gürses A, Açıkyıldız M, Güneş K, Gürses MS. Dyes and pigments: Their structure and properties. In: Dyes and Pigments. Cham: Springer Briefs in Molecular Science. Springer; 2016. DOI: 10.1007/978-3-319-33892-7_2

[8] El-Agawany NI, Abdel-Razik SAR, Abdel-Kareem MS, et al. Environmental performance of two marine algae Ulva fasciata and *Pterocladia capillacea* in the biological treatment of four reactive dyes from aqueous solutions based on fresh and dried biomass. Applied Water Science. 2023;**13**:82. DOI: 10.1007/s13201-023-01893-7

[9] Szymczyk M. Chromophore. In: Shamey R, editor. Encyclopedia of Color Science and Technology. Cham: Springer; 2023. DOI: 10.1007/978-3-030-89862-5_440

[10] Shahin F. The influence of cationization on the dyeing performance of cotton fabrics with direct dyes. Journal of Engineering Research and Applications. 2015;**5**(8, (Part – 3)):62-70, ISSN: 2248-9622

[11] Bechtold T, editor. Handbook of Natural Colorants. John Wiley & Sons Ltd: Chichester, England; 2023

[12] Calderin J. Form, Fit, Fashion. Boston, Massachusetts, United States of America: Rockport; 2009. p. 125. ISBN 978-1-59253-541-5

[13] SCI Perkin Medal. Science History Institute. 2016. Available from: https://www.sciencehistory.org/sci-perkin-medal [Accessed: March 24, 2018]

[14] Hicks Jan. William Henry Perkin and the world's first synthetic dye. Science and Industry Museum blog. 2017. Available from: https://blog.scienceandindustrymuseum.org.uk/worlds-first-synthetic-dye/

[15] Salih E, Hanan A, Ali A-S. Preparation of novel azo dyes as a new anti-human breast cancer MDA-MB231 cells and study its association with DNA. Journal of Basrah Researches (Sciences). 2019;**45**(2):22-32. Available from: http://brsj.cepsbasra.edu.iq

[16] Lipskikh OI, Korotkova EI, Khristunova Ye P, Barek J, Kratochvil B. Sensors for voltammetric determination of food azo dyes - a critical review. Electrochimica Acta. 2018;**260**:974-985

[17] 4-phenylazophenol. 2019. Available from: https://www.chemicalbook.com/ChemicalProductProperty_EN_CB5366310.htm; https://www.chemicalbook.com

[18] Benkhaya S, M'rabet S, El Harfi A. Classifications, properties, recent synthesis and applications of azo dyes. Heliyon. 2020;**6**(1):e03271, e03271. DOI: 10.1016/j.heliyon.2020, ISSN 2405-8440. Available from: https://www.sciencedirect.com/science/article/pii/S240584402030116X

[19] Siddiqua UH, Ali S, Iqbal M, Hussain T. Relationship between structure and dyeing properties of reactive dyes for cotton dyeing. Journal of Molecular Liquids. 2017;**241**:839-844. DOI: 10.1016/j.molliq.2017.04.057, ISSN 0167-7322. Available from: https://www.sciencedirect.com/science/article/pii/S0167732217305561

[20] Al-Etaibi AM, El-Apasery MA. A comprehensive review on the synthesis and versatile applications of biologically active pyridone-based disperse dyes. International Journal of Environmental Research and Public Health. 2020;**17**(13):4714. DOI: 10.3390/ijerph17134714

Chapter 10

Advancing Sustainable Approaches for the Removal and Recycling of Toxic Dyes from the Aquatic Environment

Jia-Ren Chang Chien and Janet Joshiba Ganesan

Abstract

The widespread usage of synthetic dyes and chemicals across industries leads to the production of a considerable amount of wastewater. Textile industries, in particular, frequently release harmful dyes directly into the environment, presenting significant threats to human health. Discharging untreated sewage from numerous textile industries contributes to severe environmental consequences. To address these concerns, there is growing emphasis on developing efficient and cost-effective nano-adsorbents, leveraging their distinctive properties to mitigate the ecological impact of dyes. In light of these challenges, significant efforts have been focused on developing efficient adsorbents to remove unwanted substances from wastewater. Effectively and selectively eliminating dyes is increasingly acknowledged as essential for addressing environmental concerns. A comprehensive approach involves comparing the degradation efficiency of different catalysts, considering the search for a suitable adsorbent. This chapter comprehensively examines numerous advanced adsorbents in the literature for dye degradation. In addition to outlining the physicochemical characteristics of various adsorbents, the chapter delves into the mechanisms and effectiveness of the adsorption process. It is crucial to tackle the challenges inherent in dye degradation to alleviate the adverse environmental effects of dyes. A critical evaluation of next-generation adsorbents can advance the adoption of clean and cost-effective water purification practices.

Keywords: hazardous impacts, dyes, adsorbents, textile wastewater, nanomaterials

1. Introduction

Water is Earth's most precious natural resource, but several kinds of pollution are constantly tainting it. The demand for clean water is steadily rising because of the population explosion, severely threatening our ecosystem [1]. In today's life, lack of access to sanitary facilities and clean drinking water is one of the most inescapable problems affecting people. Water pollution is a critical global issue, exacerbated by widespread industrialization, human activities, unsustainable water resource exploitation, unbridled growth of dye-utilizing industries, and inadequate effluent treatment methods [2–4]. The wastewater liberated from factories and laboratories involved in the textile industry have led to serious pollution issues because unused dyes are released into the environment, even in minimal amounts (less than 1 ppm for some specific dyes), which are hazardous due to their high toxicity and color [1, 5, 6]. The United Nations (UN) has reported that significant polluting elements are not removed from over 80% of wastewater produced by anthropogenic activities before it is released into waterways. By 2030, it is predicted that 47% of people on Earth will reside in areas with limited access to clean water [6]. In recent years, the dye industry has expanded rapidly. According to the US "Color Index," the quantity of commercial dyes has surged to tens of thousands. Annually, approximately 60,000 tons of dyes, with around 80% being azo dyes, are disposed of into the environment globally as waste [5, 7]. The paper and textile industries utilize a lot of synthetic dyes. Over 10,000 dyes and pigments are used in the textile industry, along with an ineffective dyeing water treatment procedure that poses a severe risk to human health [6]. In rivers where treated wastewater is discharged, dyes can undergo partial degradation or transformation in the water and sediment. Certain pigments and compounds they convert into are toxic carcinogens [5, 8].

Globally, there is increasing concern about eliminating microorganisms, hazardous dyes, and toxic contaminants to provide potable water. Most countries now use costly water purification systems, making them unaffordable for many low- and middle-income nations [6]. This chapter aims to present a thorough analysis of several cutting-edge sustainable approaches in wastewater treatment techniques for eliminating toxic dyes from aqueous solutions. Furthermore, a detailed method compares the effectiveness of various catalysts in degrading materials while considering the need to find an appropriate adsorbent. Moreover, this chapter comprehensively reviews various state-of-the-art adsorbents documented in the literature for dye degradation. In addition to emphasizing the various adsorbent's physicochemical characteristics, the chapter also covers the mechanisms and efficiencies associated with the adsorption process. To lessen the harmful effects of dyes on the environment, addressing the difficulties posed by the dye degradation process is essential. The features, benefits, limitations, and underlying mechanisms of various sorbents have all been thoroughly examined. The difficulties in creating a dye degradation process that is both efficient and environmentally benign are emphasized. This chapter will improve knowledge of the state-of-the-art dye degradation process and be helpful in the actual use of various sorbents for particular dye adsorption.

2. Textile dyes

The textile industry has become a primary contributor to pollution in surface and groundwater reservoirs. The main uses of dyes are to color textiles and materials such as plastics, leather, fur, and hair. In several nations, including China and the estuaries of South Africa, the textile industries significantly contribute to environmental degradation and the growth of the global economy [6]. The textile industry utilizes over 10,000 tons of synthetic dyes annually, accounting for a fraction of the approximately 7×10^7 produced globally each year [9]. Dyes can absorb or emit light within the visible range (400–700 nm), making color. Typically, a dye requires a complex aromatic conjugated system for resonance, essential for imparting color. Within a dye's aromatic rings exist color-bearing groups known as chromophores and other chemical groups such as bridging groups (auxochromes). These auxochromes intensify or modify chromophores of color and facilitate the fixation of dyes [10]. Dye is commonly classified into multiple categories based on its source, composition, and usage. Among synthetic dyes, the textile industries predominantly employ azo, direct, reactive, mordant, acid, basic, dispersion, and sulfide dyes. The textile business uses natural fibers, such as acrylic, polyester, wool, cotton, and silk [9]. Two main categories of dyes exist: synthetic and natural. Synthetic dyes are extensively utilized across various industries, offering a broader spectrum of colors than natural dyes. Synthetic dyes can be divided into two primary groups: ionic and non-ionic. There are cationic and anionic dyes within the ionic group. Even in small amounts, all of these dyes have the potential to be harmful when dissolved in water. Based on their specific applications, they can be further divided into subcategories, such as reactive, direct, essential, and dispersed dyes depicted in **Figure 1** [11]. Most synthetic dyes are sourced from two primary origins: petroleum-based intermediates and coal tar. These dyes are typically available in various forms, including liquid dispersions, pastes, granules, or powders [12]. A recent investigation found that 100,000 commercially available dyes were used globally, with an annual consumption record of 10,000 tons [13]. During the washing of colored or printed textile goods, a substantial volume of liquid waste containing dyes and pigments is generated. This waste constitutes approximately 60% of the total volume [14–16]. Most dyes are complex organic compounds that can adhere to various surfaces, including leather, textiles, and so on. The complexity of effluent and the massive dissipation of textile colors disturb the ecological system [13].

Inefficient textile dyeing techniques, where dyes fail to bind to fibers and fabrics properly, lead to the release of 15–50% of azo dyes into the wastes of textile industries [14]. Industrial effluents contaminated with dyes pose significant environmental hazards due to elevated levels of biological oxygen demand (BOD), chemical oxygen demand (COD), total dissolved solids (TDS), as well as various toxic heavy metals and harmful chemical compounds [10].

For instance, textile effluents have a high concentration of dyes and numerous other compounds added during the coloring process. The high intensity of color and wide compositional variability of textile effluent make treatment extremely challenging [4].

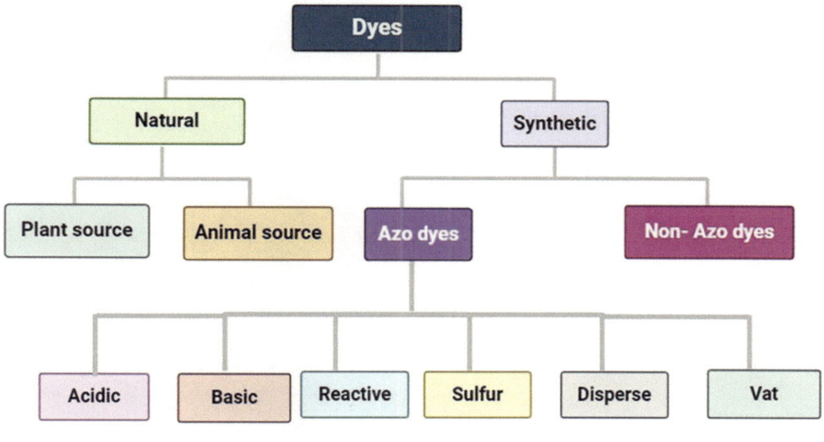

Figure 1.
Classification of dyes.

3. Environmental and health impacts of toxic dyes

Wastewater containing dyes presents substantial risks, carrying the potential for adverse effects like cancer and mutagenesis in humans if released without treatment. This risk is predominantly attributed to these dyes' complex chemical structure and non-biodegradable characteristics. Globally, an estimated annual volume of at least 200 billion liters of wastewater from dyeing is generated [17]. Given the global health risks associated with textile wastewater discharge, new initiatives for environmental restoration are being launched, driven by both ecological preservation and commercial interests [18]. The extensive usage of textile dyes and their potential to create hazardous aromatic amines make them environmentally important. Additionally, they pose a risk to ecosystems because they block sunlight, which lowers photosynthetic activity and dissolved oxygen concentration [10]. Azo dyes pose a hazard due to toxic amines in the effluents. Additionally, anthraquinone dyes maintain their color for extended periods in textile rejects owing to their strong resistance to deterioration. Numerous adverse environmental effects can be traced back to the textile industry [13]. Particulate matter, dust, sulfur and nitrogen oxides, and volatile organic compounds are a few examples of the created air pollution. The primary sources of solid waste in the textile industry include leftover yarns, fabric remnants, and empty packaging. In contrast, textile sludge presents challenges related to excessive volumes and undesirable composition, often containing high concentrations of organic matter, micronutrients, heavy metal cations, and pathogenic microorganisms [16, 17].

The consumption of azo dye through food or water can result in acute vomiting, cyanosis, allergic reactions, and genetic mutations [8]. One of the most critical consequences of textile waste is the depletion of dissolved oxygen in water. Dissolved oxygen is essential for marine life, and its depletion endangers aquatic ecosystems and

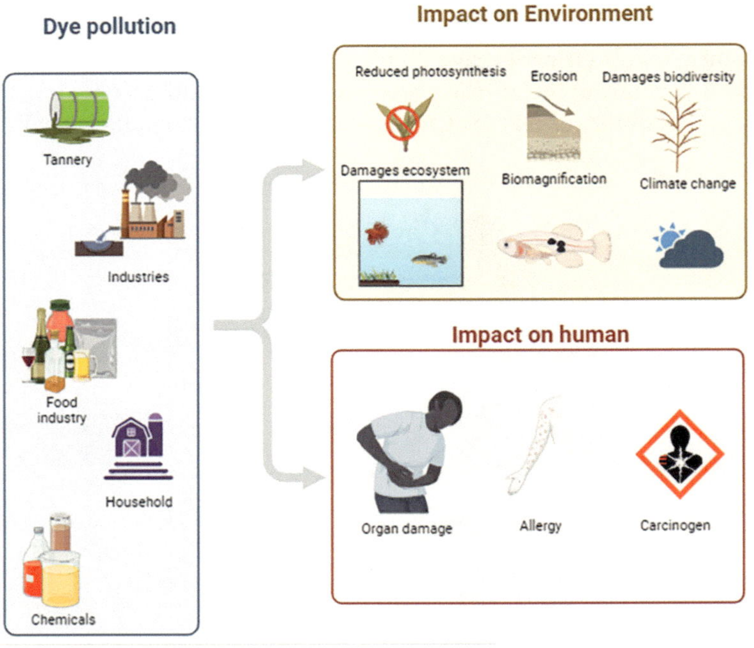

Figure 2.
Schematic representation of environmental and health impacts of dye pollution.

hinders the self-purification ability of water bodies [18]. Growing amounts of dye in water bodies impact several algae parameters, including growth protein, pigment, and other nutritional content. The possible effects of various dyes on algae vary. Algae are 50% more sensitive to pollutants when measuring contamination in aquatic habitats than species typically utilized in toxicological testing [12]. Numerous dyes harm flora and fauna, hinder the growth of microorganisms, and are hazardous to fish and mammals. In addition, several dyes and the byproducts of their breakdown are dangerous to aquatic life [18]. Even the treated water contains pigments that may be primarily destroyed or transformed in the water and sediments of streams. Both textile dyes and their byproducts have the potential to cause cancer, mutagenesis, and teratogenicity. Moreover, almost 40% of colors and dyes used worldwide contain chemically bound chlorine [19]. Along with the wastewater, these hazardous substances are transported over great distances. After that, they stay in the water and soil for extended periods, endangering living things' lives and decreasing soil fertility and aquatic plant photosynthetic activity, which causes anoxic conditions to develop for marine fauna and flora [9]. The pictorial representation of the environmental and health impacts of dye contamination is depicted in **Figure 2**.

4. Conventional treatment techniques for toxic dyes

The presence of dyes and other chemicals, in the effluent produced by the cotton dyeing and textile industries, is contaminated with high levels of color, BOD, COD, and total solids. Drinking water should be colorless and devoid of harmful substances to prevent problems with the river's treatment costs. Thus, before being released into any freshwater body, textile wastewater must undergo various treatment procedures, such as physical, chemical, and biological [20]. To address the issue of dye removal from wastewater, numerous techniques based on several methods, such as combination treatment processes and physical, chemical, and biological approaches, have been developed to treat textile dyeing wastewater economically and effectively.

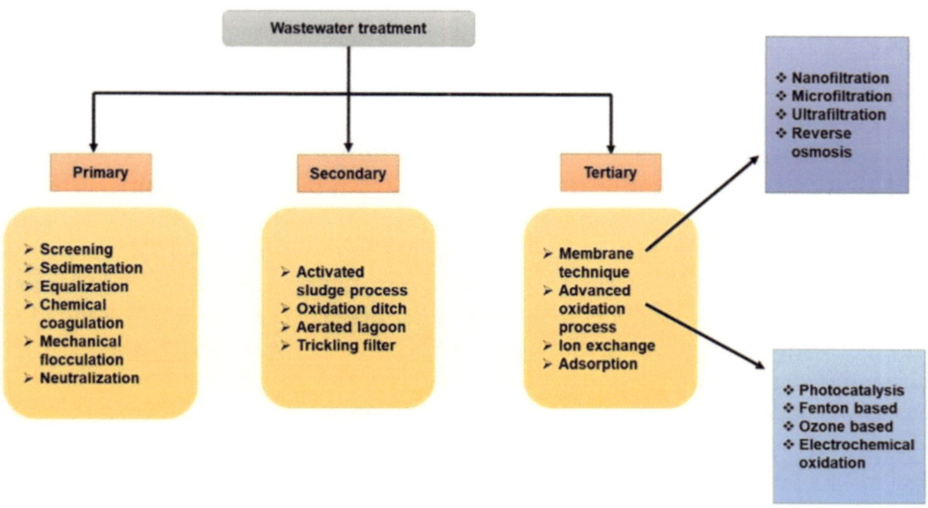

Figure 3.
Schematic representation of conventional dye wastewater treatment technologies.

These processes are typically very effective in wastewater from textile dyeing [11, 12]. The adsorption, sedimentation, ultrafiltration, flotation, coagulation, flocculation, filtration, photoionization, and incineration techniques are subdivided under the physical approaches. The methods, including neutralization, oxidation, reduction, electrochemical oxidation, wet-air oxidation, and photochemical degradation, are the subgroups of chemical techniques. Activated sludge, trickling filters, aerated lagoons, stabilization ponds, anaerobic digestion, and various microbial strains are all used in biological treatment techniques [21–23]. Compared with other conventional techniques, the adsorption technique has several benefits, such as high efficiency, rapid reaction times for wastewater color removal, and the capacity to reuse adsorbent; also, it is one of the promising techniques in sequestering toxic dyes [9]. The schematic representation of conventional dye wastewater treatment technologies is depicted in **Figure 3**.

5. Adsorption of dyes

Adsorption is a surface phenomenon where molecules or ions that have been adsorbed are drawn toward the surface of a solid adsorbent. It is divided into two types based on how the dye molecule is adsorbed onto the adsorbent surface: physisorption and chemisorption [9]. Adsorption is an eco-friendly, effective, and reasonably priced technique for sequestering contaminants from wastewater. The most popular adsorbents utilized to remove dyes and heavy metals are composites, metal oxides, polymeric polymers, and activated carbon [8, 24, 25]. The adsorption method is a desirable treatment for dye-contaminated wastewater. Because of its effectiveness and affordability, adsorption has become one of the widely used wastewater treatment techniques in industries [5]. The adsorption process is supported by various forces such as hydrogen bonding, van der Wall, hydrophobic attraction, and electrostatic forces. The basic principle of adsorption is that the porous structure present on the surface of the adsorbent enhances the adsorption in the total exposed surface area necessary for the quick and effective adsorption of dye molecules from wastewater [9]. The adsorbent may be recycled and regenerated simultaneously without producing further pollution. As a result, adsorption is a separation technology with several uses. The most crucial phase in adsorption is preparing superior adsorbents that can efficiently remove the dye [26]. Adsorption is one of those methods that can help remove organic chemicals that are difficult for the body to break down effectively. Because of their broad surface area and porous nature, typical activated carbons (ACs) effectively remove the organic molecules [27]. A biosorbent significantly aids the elimination of dye. Utilizing household and agricultural wastes as adsorbents has become a practical substitute. Many adsorbents from biomass waste have been developed and are used as very efficient agents to remove contaminants from wastewater and water. These waste products were used either precisely as they were intended to be used or with the appropriate changes. Several agricultural and food waste materials have shown promise when used as adsorbents to remove contaminants. These materials include Azolla, banana peel, cabbage waste, chitosan, citrus peel, *Citrus limonum* leaves, corn cob, orange peel, peanut hull, rice husk, sawdust, and sugar cane bagasse [28]. Adsorbents of all kinds, including hydrogels, metal-organic frameworks, organic polymers, nanomaterials, and others, have been effectively developed for the remediation of wastewater containing dyes [29, 30]. The morphological characteristics such as size, shape, and structure of the adsorbent strongly influence the adsorption mechanism, and the capability of the adsorbent's chemical structure is incredibly crucial. As a result, adsorbents for removing pollutants have been prepared using a large amount of natural and mineral resources [31].

5.1 Carbon-based adsorbents

Biochar is a resultant product of pyrolysis of forestry and agricultural waste with many surfaces, chemical functional groups, and a porous structure. Adsorbing organic dyes from wastewater using biochar is a promising sustainable adsorbent that aligns with sustainable development and cleaner production. Regarding removing dye, biochar's adsorption efficiency is inferior to typical adsorbents and must be significantly enhanced throughout pyrolysis [32, 33]. Carbonaceous materials have become the most widely utilized adsorbents because of their many benefits, including their well-developed pore structure, exceptional physical and chemical stability, and ease of functionalization. However, conventional porous carbon materials often have a low adsorption capacity. They are challenging to recover from, severely restricting their practical use and encouraging the development of effective porous carbon adsorbents [34]. In many diverse treatment applications, such as industrial wastewater treatment from the textile, sugar, leather, pharmaceutical, and petroleum industries, activated carbons are frequently employed. Various surface functional groups, such as quinones, carbonyl, and carboxyl, can absorb pollutants from wastewater contamination. According to published research, activated carbon has a very effective capacity for adsorbing various contaminants, depending on the adsorbate and functional groups of the adsorbent. Altering the activated carbon's surface is an additional method that successfully displays alterations in the configuration of different surface functional groups [35]. Many studies have been conducted on the adsorption of dyes using carbon-based materials, such as activated carbon (AC), carbon quantum dots (CQDs), carbon nitride (C_3N_4, CN), and graphene oxide (GO) due to their high physical and chemical stability, thermal stability, and high adsorption capacity [36]. Due to these modifications, the adsorbent's surface area increased, and a new functional group and texture were created by adding atoms. Adsorption of contaminants from wastewater is substantially improved by altering physical and chemical activation processes, mainly when removing heavy metal ions and industrial colors [35]. In the work conducted by Fan et al., it is reported that through free radical polymerization of a new class of polyacrylic acid grafted lignosulfonate (SLS)/carboxymethyl cellulose (CMC) and biochar (BC) composites (SLS/CMC/BC), which can efficiently adsorb Pb^{2+} and methylene blue (MB). The highest adsorption capacity of MB and Pb^{2+} in a monopollutant system is 113 mg g^{-1} and 204 mg g^{-1}, respectively. Furthermore, adding biomass carbon significantly enhanced the hydrogel's stability and adsorption capacity [26]. Through pyrolysis, activated carbon (AC) from watermelon shells was converted into the unique magnetic nanocomposite activated carbon (MWMSC). The experimental data were tested using adsorption isotherms, and the results are well-fitted to the Langmuir model, with MB and MO having maximal adsorption capacities of 303.30 and 345.70 mg g^{-1}, respectively [37]. The present understanding of techniques such as pyrolysis, physical activation, and chemical activation processes helps develop activated carbon (AC) and biochar from organically sourced materials, previously living materials considered waste materials and would otherwise be conventionally disposed of Medeiros et al. [27]. Numerous studies have been conducted to increase the adsorption capacity of activated carbon adsorbents; these studies have primarily focused on altering process parameters, such as raw material or adsorbent, carbonization, and activation procedure, under the right conditions [35]. A carbon-based adsorbent was created by carbonizing the pitaya peel (*Hylocereus undatus*) in the presence of $ZnCl_2$ to remove the Metanil Yellow dye (MY) from colored fluids. At 298 K, the Langmuir model reached its maximum adsorption capacity of 144.07 mg g^{-1}, demonstrating the best adjustment of the system's equilibrium isotherms [38].

5.2 Nanotube and graphene-based adsorbents

Carbon nanotubes (CNTs) are composed of graphene or graphite sheets coiled into tubular shapes. Their lengths are in the micrometer range, and their interior diameters are in the nanometer range. Single and multiwalled carbon nanotubes are distinguished from one another by their layered structure [39]. Zhang et al. have investigated the adsorption activity of mesoporous carbon nanospheres for the sequestration of toxic dyes such as Congo red (CR), malachite green (MG), and methylene blue (MB). The research findings concluded that the distinct porosity structure of mesoporous carbon nanospheres gives them excellent adsorption powers for dyes, such as CR (1726.60 mg g^{-1}), MG (980.55 mg g^{-1}), and MB (708.82 mg g^{-1}), indicating that they can adsorb both anionic and cationic dyes [34]. Since CNTs are considered a superior adsorption material, several studies have been conducted on their ability to remove contaminants from aqueous systems. Their high specific area, high porosity, large ratio of accessible micropores, hollow internal structure, layered structure, appropriately established mesopores, hydrophobicity, distinct sidewall curvature, Pi-conjugative structure, ease of chemical activation, lightweight, tunable surface properties, high corrosion resistance, excellent support for coating or depositing other materials, and high water permeability make them exceptional adsorbents [39]. Researchers are drawn to carbon-based materials among various inorganic substances, such as graphene oxide (GO) and carbon nanotubes (MWCNTs), due to their rich specific surface area and abundance of chemical modification sites [29]. A straightforward ultrasound irradiation technique was used to create a variety of carbon-based $MgAl_2O_4$ adsorbents using activated carbon (AC), carbon quantum dots (CQDs), C_3N_4 (CN), and graphene oxide (GO) as carbon sources. The Congo red (CR) dye was adsorbed using carbon-based MAO adsorbents, which produced an 89.7% dye removal rate [36].

5.3 Metal-based adsorbents

Metal-organic frameworks (MOFs) and materials generated from MOFs are becoming more and more attractive as adsorption candidates because of their essential characteristics, which include a large surface area, adjustable porosity, and a variety of functions [40]. Metal oxide nanoparticles such as ZnO, TiO_2, $BiVO_4$, and WO_3 have drawn increased interest because of their potential uses in solar fuel, energy conversion and storage, and photocatalysis. Moreover, metal oxides are employed in various chemical redox processes, including eliminating dye contamination from wastewater [41]. A recently discovered class of two-dimensional (2D) transition metal carbides, nitrides, or carbonitrides known as MXenes has shown promise in several domains. Because of their graphene-like morphology and distinctive qualities such as their large specific surface area, highly active sites, hydrophilic nature, environmental friendliness, thermal and electrical conductivity, anisotropy, anti-microbial properties, large inter-layer spacing, high chemical stability, superior sorption/reduction capacity, and high negative zeta-potential—these materials can be used in a variety of applications. Because of their layered structure, high hydrophilicity, wide specific surface area, and negatively charged surface, MXenes are the perfect adsorbent materials to adsorb dyes from the aqueous environment [42]. One notable adsorbent that is affordable, simple to synthesize, and non-toxic to human health is magnesium oxide (MgO). This work used a straightforward sol-gel process to manufacture magnesium oxide, which is inexpensive and environmentally benign. For the first time, MgO microparticles were used to remove Reactive Red 21 azo dye, and their effectiveness in actual textile effluent was examined. The adsorption studies inferred that the MgO adsorbent effectively sequestered about 98% of the azo dye in real-time textile effluent [43]. The ZnO/MgO nanocomposite was made using the Pechini sol-gel method, with tartaric acid as a chelating agent and ethylene glycol as a crosslinker. A ZnO/MgO nanocomposite was employed to remove Congo red dye from aqueous solutions rapidly. The ZnO/MgO nanocomposite has a maximal Congo red dye adsorption capability of 317.46 mg g^{-1} [44]. On the other hand, the inability of MOFs to withstand severe environmental conditions led to the development of more stable MOF-derived materials for adsorption purposes. A growingly effective technique for creating stable metal oxide-carbon hybrids as adsorbents for removing pollutants is the carbonization of MOFs [40]. Metal-organic porous frameworks with large surface area, adjustable porosity, and usefulness have garnered much interest in adsorption. Because of their distinctive structural characteristics, MOFs have the potential to be used as materials for a variety of other purposes, such as medication delivery, gas storage, sensors, heterogeneous catalysis, and separation. MIL-100(Fe)-derived mesoporous $LaFeO_3$@C nanocomposites for the efficient adsorption removal of sulforhodamine B (SRB) and methyl orange (MO) were studied by Lalan et al. Regarding SRB and MO, the material formed at 700°C (LFO@C-700) showed greater removal effectiveness of 99% and 97.1%, respectively [40].

5.4 Polymer and gel-based adsorbent

The wastewater treatment industry extensively uses polymer materials as functional materials because of their distinctive structure, ease of modification, and processing. Polymeric materials offer a wide range of pore diameters, chemical stability under a variety of environmental circumstances, and affinity for target dyes based on the functional groups of the polymer as well as the dyes; they are a viable candidate for the adsorption of dyes from aqueous fluids [26]. Recently, several natural polymers, including gelatin, cellulose, polyvinyl alcohol, chitosan, and sodium alginate, as well as mineral minerals like clay, nano-hydroxyapatite, calcium oxide, and zinc, have been used to synthesize adsorbents because of their cost-effectiveness and efficacy [31]. In the study conducted by Zhou et al., an efficient and unique porous alginate/gelatin/n-hydroxyapatite/magnetic nanoparticle was fabricated and utilized to sequester methylene blue [31]. In the work conducted by Brahmi et al., porous hydroxyapatite-metakaolin geopolymer (HAP-MK-GP) granules are used as a unique adsorbent for capturing bright green dye. Metakaolin, solid sodium metasilicate, and finely ground hydroxyapatite were combined in a high-shear granulator for the preparation. According to the findings, it is inferred that the HAP-MK-GP granules are a workable adsorbent for the removal of color from industrial effluents [45]. Researchers have focused on hydrogels, among other adsorbent materials, because of their abundance of pore architectures, wealthy functional groups, ease of synthesis and separation, and other benefits. As a result, using hydrogel adsorbent to remove organic dyes from wastewater is extremely promising. When hydrogels are formed, the insertion of inorganic compounds is frequently preferred due to the unique functionality and structural stability of hydrogels [29]. Preetha and Visalakshi fabricated an effective gel adsorbent material by grafting Karaya gum with N, N′-dimethyl acrylamide, which showed strong adsorption properties for cationic dyes. The gel is economical and environmentally friendly because it utilizes natural polysaccharides. The synthetic material exhibited swelling behavior that was pH-responsive [46] in this study conducted by Li and colleagues; graphene oxide (GO), carbon nanotubes (MWCNTs), diethylenetriamine pentaacetic acid (DTPA), acrylic acid (AA), N-vinyl imidazole (NVI), acryloyl chloride modified β-cyclodextrin (A-β-CD) as raw materials were first used to synthesize a novel and environmentally friendly pH-responsive carbon-based hydrogel adsorbent [29]. Joshiba et al. investigated the noxious Rhodamine B dye sequestration from an aqueous system using an ecologically friendly $Fe_3O_4@SiO_2$ nanocomposite that was immobilized with *Pseudomonas fluorescens* biomass in calcium alginate beads (MSAB). The research findings concluded that Rhodamine B was sequestered with a Langmuir monolayer adsorption capacity of about 229.6 mg g^{-1} [47]. Song et al. investigated methyl orange sequestration using an amine-rich gel (ARG) adsorbent derived from peach gum polysaccharide. The results showed that the ARG showed superior adsorption capacity and selectivity for anionic dyes. ARG can uptake methyl orange up to 1949.5 mg g^{-1} [48]. Li et al. created an effective sodium alginate gel adsorbent with a highly porous nature and adsorption capability. In the presence of cations, specifically Ca^{2+}, sodium alginate underwent fast cross-linking, forming a very stable and porous gel bead adsorbent (CPS-Bead). During the gel bead manufacturing process, a 3D composite carbon nanomaterial consisting of graphene oxide and carbon nanotubes modified with poly-ethylenimine (PEI) was used as a modifier of sodium alginate to enhance the adsorbent's adsorption ability further. The number of adsorption sites in adsorbents was increased by including 3D carbon materials [49]. Okara biochar was produced by pyrolysis and then further processed into the shape of beads to prevent biochar from leaking into the surrounding environment. The findings demonstrate that OBB has an adsorption percentage of up to 80% and may be readily isolated from water [30].

5.5 Photocatalyst-based adsorbents

Photocatalytic degradation employs a plentiful resource that can directly harness solar radiation. Adsorption with photocatalysis degradation techniques will be a successful strategy for managing azo-dye contaminants. The contaminants that have been adsorbed are continually photodegraded and can be eliminated because of the synergistic impact of adsorption and photodegradation. The binding sites on the materials will be liberated to allow the pollutants' continuous adsorption as the pollutants degrade concurrently. Therefore, the limitations of conventional adsorbents can be addressed, and the treatment impact of contaminants can be improved by developing materials that have the synergistic effects of adsorption and photocatalytic degradation [50]. Zero-dimensional carbon nanomaterials have demonstrated great promise as photocatalysts in the treatment of water purification as adsorbents and ion detectors and in the remediation of hazardous contaminants due to their ultra-small size, exceptional properties, and antibacterial action [51]. Heterojunction hydrogen and sodium titanate nanosheets are fabricated using microwave-assisted hydrothermal, which is easy and economical to synthesize. The synthesized material was checked for the ability of the materials to decolorize Rhodamine B, and methylene blue dyes were investigated for photocatalytic effectiveness. With a half-life period of only 5 minutes for RhB discoloration and a high adsorption capacity of around 90% in 10 minutes in the MB medium, the photocatalyst exhibiting the highest efficiency was found [52]. Rhodamine B (RB) and malachite green (MG) removal from wastewater is achieved with a one-pot approach employing microwave heating to create a biochar-based magnetic photocatalyst (ZnFe-BC). According to the Langmuir model, the adsorption capacities of MG and RB are 576.73 mg g^{-1} and 334.89 mg g^{-1}, respectively [32]. In the study conducted by Yang et al., ZnS/$CuFe_2O_4$/MXene (ZSCFOM) composite with ternary heterostructures was utilized in photodegrading the azo dyes. With the support of electrostatic interactions and hydrogen bonding, the azo dyes, such as direct brown M and direct black RN, have adsorption capacities of about 377 mg g^{-1} and 390 mg g^{-1}, respectively [50]. Cheng et al. reported that the ZnFe-BC synthesized in a single pot method utilizing microwave heating is used in effectively sequestering to remove malachite green and Rhodamine from wastewater. The research findings demonstrated the potential of ZnFe-BC as a magnetic adsorbent-photocatalyst for extracting malachite green and Rhodamine from sewage. According to the Langmuir model, the maximum adsorption quantities of malachite green and Rhodamine are 576.73 mg g^{-1} and 334.89 mg g^{-1}, respectively [32].

6. Conclusion and future perspectives

Water is the most fundamental element for all living things. In recent years, several countries have faced widespread contamination of natural water resources by organic and inorganic pollutants. It is estimated that over 800 million people worldwide still lack access to safe drinking water suitable for residential use. Textile effluent contains a broad spectrum of hazardous xenobiotics that are dangerous to the environment and public health. The daily release of millions of gallons of highly polluted wastewater by textile mills worldwide presents a significant issue with toxic dye-containing wastewater. Textile dyes are stable and can linger in the environment long if not adequately treated. However, because there is yet to be a single, practical method for effectively handling this kind of issue, treating wastewater that contains dyes is a significant challenge. Numerous established and novel methods of treating wastewater containing dyes have been documented. Adsorption is a cost-effective, efficient, simple, and eco-friendly technique compared to conventional treatment technologies. The selection of absorbent material is challenging due to various factors, including cost-effectiveness, operational simplicity, reusability, safety, and environmental impact. While low-cost carbon materials have shown efficiency in wastewater decolorization, their separation from water post-reaction poses difficulties and incurs high costs. Activated carbon is the most commonly used adsorbent for reducing wastewater pollution, but it faces significant constraints due to high regeneration and synthesis costs. Additionally, the pH sensitivity of these materials limits their effectiveness in reducing a wide variety of contaminants in water. Carbon nanotubes (CNTs) are another prominent nanomaterial for water purification, capable of removing both organic and inorganic pollutants as catalysts. However, their long-term stability is reduced due to oxidation during reactions, affecting reproducibility. Graphene-based catalysts show promise in meeting commercial demands for wastewater treatment due to their high surface area and efficient reduction of various contaminants, but they suffer from high synthesis costs. Coordination polymers have also shown significant pollutant removal from water by anchoring organic or inorganic ligands onto heavy metal complexes, though mass production remains a challenge. Harvesting atmospheric pure water presents a novel strategy for water purification. Photocatalytic dye degradation relies on factors such as light intensity and sensitization of dyes, with a need for adjustments in energy band gaps and cost reduction of photocatalysts for commercially applicable processes. Despite the laboratory-scale success, challenges in engineering, economics, and environmental factors must be addressed for widespread application. Understanding the photochemical stability under natural conditions is crucial for evaluating the effect of surface functions on dye degradation. It appears unrealistic to rely on a single globally applicable adsorbent as an end-of-pipe solution. Instead, a combination of suitable adsorbent techniques is imperative to develop technically and economically feasible options.

Author details

Jia-Ren Chang Chien* and Janet Joshiba Ganesan
Railway Technical Centre, National Kaohsiung University of Science and Technology, First Campus, Kaohsiung City, Taiwan

*Address all correspondence to: ccjr@nkust.edu.tw

References

[1] Shabir M et al. A review on recent advances in the treatment of dye-polluted wastewater. Journal of Industrial and Engineering Chemistry. 2022;**112**:1-19

[2] Mashkoor F, Nasar A. Environmental application of agro-waste derived materials for the treatment of dye-polluted water: A review. Current Analytical Chemistry. 2021;**17**(7):904-916

[3] Boukarma L et al. Novel insights into crystal violet dye adsorption onto various macroalgae: Comparative study, recyclability and overview of chromium (VI) removal. Bioresource Technology. 2024;**394**:130197

[4] Benjelloun M et al. Recent advances in adsorption kinetic models: Their application to dye types. Arabian Journal of Chemistry. 2021;**14**(4):103031

[5] Olakunle MO et al. Combating dye pollution using cocoa pod husks: A sustainable approach. International Journal of Sustainable Engineering. 2018;**11**(1):4-15

[6] Islam A et al. Step towards the sustainable toxic dyes removal and recycling from aqueous solution—A comprehensive review. Resources, Conservation and Recycling. 2021;**175**:105849

[7] Liu Q. Pollution, and treatment of dye waste-water. IOP Conference Series: Earth and Environmental Science. 2020;**514**(5):052001

[8] Anastopoulos I et al. A comprehensive review on adsorption of reactive red 120 dye using various adsorbents. Journal of Molecular Liquids. 2024;**394**:123719

[9] Al-Tohamy R et al. A critical review on the treatment of dye-containing wastewater: Ecotoxicological and health concerns of textile dyes and possible remediation approaches for environmental safety. Ecotoxicology and Environmental Safety. 2022;**231**:113160

[10] Herath IS et al. Textile dye decolorization by white rot fungi—A review. Bioresource Technology Reports. 2024;**25**:101687

[11] Tolkou AK et al. Simultaneous removal of anionic and cationic dyes on quaternary mixtures by adsorption onto banana, orange and pomegranate peels. Colloids and Surfaces A: Physicochemical and Engineering Aspects. 2024;**685**:133176

[12] Samchetshabam Gita Samchetshabam G, Ajmal HAH, Choudhury TG. Impact of textile dyes waste on aquatic environments and its treatment. Environment and Ecology;**35**(3C):2349-2353

[13] Bensalah J. Removal of the textile dyes by a resin adsorbent polymeric: Insight into optimization, kinetics and isotherms adsorption phenomenally. Inorganic Chemistry Communications. 2024;**161**:111975

[14] Dutta S et al. Contamination of textile dyes in aquatic environment: Adverse impacts on aquatic ecosystem and human health, and its management using bioremediation. Journal of Environmental Management. 2024;**353**:120103

[15] Manzoor J, Sharma M. Impact of textile dyes on human health and environment. In: Wani KA, Jangid NK, Bhat AR, editors. Impact of Textile Dyes

on Public Health and the Environment. Hershey, PA, USA: IGI Global; 2020. pp. 162-169

[16] Slama HB et al. Diversity of synthetic dyes from textile industries, discharge impacts and treatment methods. Applied Sciences. 2021;**11**:6255. DOI: 10.3390/app11146255

[17] Lellis B et al. Effects of textile dyes on health and the environment and bioremediation potential of living organisms. Biotechnology Research and Innovation. 2019;**3**(2):275-290

[18] Khan S. Malik A. Environmental and health effects of textile industry wastewater. In: Malik A, Grohmann E, Akhtar R (editors). Environmental Deterioration and Human Health: Natural and Anthropogenic Determinants. Dordrecht Netherlands: Springer; 2014. pp. 55-71

[19] Islam T et al. Impact of textile dyes on health and ecosystem: A review of structure, causes, and potential solutions. Environmental Science and Pollution Research. 2023;**30**(4):9207-9242

[20] Mani S, Chowdhary P, Bharagava RN. Textile wastewater dyes: Toxicity profile and treatment approaches. In: Bharagava RN, Chowdhary P, editors. Emerging and Eco-Friendly Approaches for Waste Management. Singapore: Springer Singapore; 2019. pp. 219-244

[21] Alsukaibi AKD. Various approaches for the detoxification of toxic dyes in wastewater. Processes. 2022;**10**:1968. DOI: 10.3390/pr10101968

[22] Marimuthu S et al. Silver nanoparticles in dye effluent treatment: A review on synthesis, treatment methods, mechanisms, photocatalytic degradation, toxic effects and mitigation of toxicity. Journal of Photochemistry and Photobiology B: Biology. 2020;**205**:111823

[23] Singh K, Arora S. Removal of synthetic textile dyes from wastewaters: A critical review on present treatment technologies. Critical Reviews in Environmental Science and Technology. 2011;**41**(9):807-878

[24] Zaferani SPG, Amiri MK, Amooey AA. Computational AI to predict and optimize the relationship between dye removal efficiency and Gibbs free energy in the adsorption process utilizing TiO_2/chitosan-polyacrylamide composite. International Journal of Biological Macromolecules. 2024;**264**:130738

[25] Teo SH et al. Sustainable toxic dyes removal with advanced materials for clean water production: A comprehensive review. Journal of Cleaner Production. 2022;**332**:130039

[26] Fan X et al. Efficient capture of lead ion and methylene blue by functionalized biomass carbon-based adsorbent for wastewater treatment. Industrial Crops and Products. 2022;**183**:114966

[27] da Silva Medeiros DCC et al. Review on carbon-based adsorbents from organic feedstocks for removal of organic contaminants from oil and gas industry process water: Production, adsorption performance and research gaps. Journal of Environmental Management. 2022;**320**:115739

[28] Periyasamy AP. Recent advances in the remediation of textile-dye-containing wastewater: Prioritizing human health and sustainable wastewater treatment. Sustainability. 2024;**16**:495. DOI: 10.3390/su16020495

[29] Li X, Li K. Multifunctional pH-responsive carbon-based hydrogel

adsorbent for ultrahigh capture of anionic and cationic dyes in wastewater. Journal of Hazardous Materials. 2023;**449**:131045

[30] Zhou G et al. A new type of highly efficient fir sawdust-based super adsorbent: Remove cationic dyes from wastewater. Surfaces and Interfaces. 2023;**36**:102637

[31] Zhou W et al. Synthesis and characterization of Alg/Gel/n-HAP/MNPs porous nanocomposite adsorbent for efficient water conservancy and removal of methylene blue in aqueous environments: Kinetic modeling and artificial neural network predictions. Journal of Environmental Management. 2024;**349**:119446

[32] Cheng S et al. Preparation of magnetic adsorbent-photocatalyst composites for dye removal by synergistic effect of adsorption and photocatalysis. Journal of Cleaner Production. 2022;**348**:131301

[33] Suratman A et al. Okara biochar immobilized calcium-alginate beads as eosin yellow dye adsorbent. Results in Chemistry. 2024;7:101268

[34] Zhang X et al. A novel mesoporous carbon nanospheres-based adsorbent material with desirable performances for dyes removal. Journal of Molecular Liquids. 2023;**390**:123091

[35] Raninga M et al. Modification of activated carbon-based adsorbent for removal of industrial dyes and heavy metals: A review. Materials Today Proceedings. 2023;**77**:286-294

[36] Wang S et al. Various carbon-based $MgAl_2O_4$ adsorbents and their removal efficiency of CR dye and antibiotics in aqueous media: High selective adsorption capacity, performance prediction and mechanism insight. Ceramics International. 2023;**49**(16):26734-26746

[37] Rajendran J et al. Methylene blue and methyl orange removal from wastewater by magnetic adsorbent based on activated carbon synthesised from watermelon shell. Desalination and Water Treatment. 2024;**317**:100040

[38] Georgin J et al. Residual peel of pitaya fruit (*Hylocereus undatus*) as a precursor to obtaining an efficient carbon-based adsorbent for the removal of metanil yellow dye from water. Journal of Environmental Chemical Engineering. 2022;**10**(1):107006

[39] Sajid M et al. Carbon nanotubes-based adsorbents: Properties, functionalization, interaction mechanisms, and applications in water purification. Journal of Water Process Engineering. 2022;**47**:102815

[40] Lalan V et al. Metal-organic framework-derived $LaFeO_3$@C: An adsorbent for removing organic dyes from water. Journal of Environmental Chemical Engineering. 2023;**11**(6):111405

[41] Hemmatzadeh E, Bahram M, Dadashi R. Photochemical modification of tea waste by tungsten oxide nanoparticle as a novel, low-cost and green photocatalyst for degradation of dye pollutant. Spectrochimica Acta Part A: Molecular and Biomolecular Spectroscopy. 2024;**313**:124104

[42] Bilal M, Khan U, Ihsanullah I. MXenes: The emerging adsorbents for the removal of dyes from water. Journal of Molecular Liquids. 2023;**385**:122377

[43] Deniz S. Efficient and environmentally friendly removal of azo textile dye using a low-cost adsorbent: Kinetic and reuse studies with application to textile effluent. Materials Today Communications. 2023;**35**:106433

[44] Alghanmi RM, Abdelrahman EA. Simple production and characterization of ZnO/MgO nanocomposite as a highly effective adsorbent for eliminating Congo red dye from water-based solutions. Inorganic Chemistry Communications. 2024;**161**:112137

[45] Brahmi A et al. Porous metakaolin geopolymer as a reactive binder for hydroxyapatite adsorbent granules in dye removal. Hybrid Advances. 2024;**5**:100134

[46] Preetha BK, Vishalakshi B. Karaya gum-graft-poly(N,N'-dimethylacrylamide) gel: A pH responsive potential adsorbent for sequestration of cationic dyes. Journal of Environmental Chemical Engineering. 2020;**8**(2):103608

[47] Joshiba GJ et al. Investigation of magnetic silica nanocomposite immobilized *Pseudomonas fluorescens* as a biosorbent for the effective sequestration of Rhodamine B from aqueous systems. Environmental Pollution. 2021;**269**:116173

[48] Song Y et al. Superior amine-rich gel adsorbent from peach gum polysaccharide for highly efficient removal of anionic dyes. Carbohydrate Polymers. 2018;**199**:178-185

[49] Li X et al. Facile preparation of sodium alginate gel beads enhanced by polyamino-modified 3D carbon for efficient remediation of organic dyes in wastewater. Separation and Purification Technology. 2024;**339**:126637

[50] Yang H et al. ZnS/$CuFe_2O_4$/MXene ternary heterostructure photocatalyst for efficient adsorption and photocatalytic degradation of azo dyes under visible light: Synergistic effect, mechanism, and application. Chemosphere. 2023;**339**:139797

[51] Budimir MD, Prekodravac JR. 10—Photocatalytic properties of zero-dimensional carbon–based nanomaterials: Application as catalysts/adsorbents in water treatment. In: Joseph K et al., editors. Zero-Dimensional Carbon Nanomaterials. Woodhead Publishing; 2024. pp. 291-355

[52] Pereira CAM et al. Effect of chemical potential on the structural modification of titanate-based photocatalysts: Fast dye degradation efficiency and adsorption power. Journal of Alloys and Compounds. 2023;**947**:169691